The Unseen Coastline: Exploring the Ecology of Intermediate Shores

Fedor

Table of Contents

Chapter 1: Introduction and motivation

The intertidal zone can be comprised of various shore types ranging in particle size from fine muds though to solid rock and also including various specialised habitats, such as intertidal pools, mangrove forests and wave-beaten cliffs (Raffaelli & Hawkins, 1996). To date, most intertidal research has focused on shores from the two ends of this particle size spectrum, rocky, and fine sandy/muddy shores (Raffaelli & Hawkins, 1996). To a large extent this is probably due to the fact that the vast majority of the coastline is composed of one of these two dominant habitat types, at least in this region (Griffiths et al., 2010). The biota of both shore types is also relatively easy to sample quantitatively, although by using different techniques (primarily photographic or scraped quadrats on rocky shores and cored or dug and sieved samples on soft shores). Shores of intermediate particle grain sizes are both rare and harder to sample, as the biota is usually mobile, 3-dimensional and cryptic.

Excluding estuaries, reviews on the marine biodiversity of the South African coastline (Brown & Jarman, 1978; Field & Griffiths, 1991; Bustamante & Branch, 1996; Awad et al., 2002; Griffiths et al., 2010; Scott et al., 2012; Branch & Branch, 2018) or of regional sections of that coastline, such as the Western (Jackson, 1976; Branch & Griffiths, 1988), Eastern (Griffiths, 2005) or Southern (Wooldridge, et al., 1981) part of the country make little mention of shore types other than sandy or rocky shores. Instead South Africa's shoreline has been categorized as 42% sandy, 27% rocky and 31% mixed shores (Griffiths et al., 2010) the latter category including all six types of intermixed rock and sand recognized by Bally (1981; 1987) and contain no mentions of other particle size shores. As such intertidal research in South Africa mostly concentrates on the overall biological and physical features of either rocky (Stephenson & Stephenson, 1972; Branch, 2001; Menge & Branch, 2001) or sandy (McLachlan et al., 1981c;

McLachlan & Bates, 1983; Nel, 2000; Defeo & McLachlan, 2013; McLachlan & Defeo, 2017) shores, or by studying sections of coastline throughout specific bioregions, such as Namaqua (Day, 1959; Brown, 1971) and Natal (Dye et al., 1981; Wooldridge et al., 1981). Further studies report biological patterns in overall sandy shore communities (Defeo & McLachlan, 2005; McLachlan & Dorvlo, 2005) and the effects physical environments have on them (McLachlan et al., 1993), as well as identifying sandy shore (Brown & McLachlan, 2002; McLachlan & Brown, 2006; Schlacher et al., 2007; Schlacher et al., 2008; Defeo et al., 2009; Harris et al, 2011; Harris et al., 2014; Harris et al., in press) and rocky shore (Blamey & Branch, 2009) conservational needs. More specific studies focus on food relations (Brown, 1964), energy flow (McLachlan, 1977c; McLachlan et al., 1981a), and survival behaviours (Brown, 1996) within sandy communities; and the effects of wave exposure (McLachlan et al., 1981b; McQuaid & Branch, 1984; McQuaid & Branch, 1985), recolonization (Dye, 1988) alongside Connell's (1978) Intermediate Disturbance Hypotheses (Sousa, 1985; Lasiak & Field, 1995), and temperature (McQuaid & Branch, 1984; Huggett & Griffiths, 1986) on rocky communities. Apart from a single study on boulder shores (Tucker et al., 2017), no South African studies examine communities living on shores of other particle sizes.

Thus, knowledge on shore types not categorized as either rocky or sandy is severely lacking. As most sandy shore research has been conducted on shores with mean particle sizes of 1mm or smaller and most rocky shore research on solid rock platforms, the understudied shore types encompass the wide range of shores with particle sizes in-between these groups. In this dissertation I refer to these understudied shore types collectively as the 'Missing Middle' and use particle size to subdivide these into individual recognized shore types, about which further details are provided below.

According to Wentworth's (1922) system, which classifies particles according to their size, the 'Missing Middle' shore types are identified as follows: very coarse sand 1 —< 2mm, granules 2 —< 4mm, pebbles 4 —< 64mm, cobbles 64 —< 256mm, and boulders with a particle >256mm.

The locations of some Missing Middle shores were mapped by Jackson and Lipschitz (1984) in their *'Coastal Sensitivity Atlas of Southern Africa 1984'* which recognized coarse sand and pebble/shingle shores as occurring throughout the coasts of South Africa and it's four distinct bioregions (Sink et al., 2012). Few pebble/shingle shores (<10) are shown in the Namaqua Bioregion, (West coast), increasing to ~20 pebble/shingle shores in the Agulhas Bioregion (South Coast) (Jackson & Lipschitz, 1984). Pebble/shingle shores become more numerous throughout the Natal Bioregion (East coast), which is predominately made up of coarse sand shores, with the Delagoa Bioregion (North East coast) being all coarse sand shores with small interruptions of wavecut rocky platforms (<20) towards the Mozambican border (Jackson & Lipschitz, 1984). Since Jackson and Lipschitz (1984), apart from boulder shores (Sink et al., 2012), other Missing Middle shores have not been identified or quantified along the South African coast.

Research Question and Aims

As detailed above, no ecological research has been conducted on shores comprised of particles of 2mm grain size through to cobbles (64 —< 256mm) in South Africa, and only one study has been done on boulder shores (Tucker et al., 2017).

The purpose of this dissertation is to explore the macrofauna and macroflora occurring across a full range of particle sizes on shores in the South Western Cape region of South Africa. The specific aims are as follows:

1. To collate and compare previously existing and current survey data on the composition of the biota across the full particle spectrum of particle sizes from fine sand to rocky shores to determine how many distinct habitat types occur across this particle spectrum of shores and the degree of dissimilarity between each of these habitat types.

2. Identify which taxa, or groups of taxa, best characterise each of these habitat types. This analysis forms Chapter 2 of the dissertation.

3. To explore in more detail the composition and zonation patterns of the fauna occupying pebble and cobble shores, since this group of shores has not been the subject of any faunal analyses to date. This forms Chapter 3 of the dissertation.

These two chapters are presented in the form of independent papers with their own introductions and methods sections, ready for submission for publication, hence there is some inevitable duplication in content between these sections.

Chapter 2:

Types of particulate shores existing on the South African Western Cape coast

Introduction

Most intertidal research has focused on either rocky, or sandy or muddy shores (Raffaelli & Hawkins, 1996) with extremely little information available on the biota of any shore type not at these two ends of the particle-size spectrum. According to the Wentworth (1922) scale, most sandy shores have a mean grain size of 1mm or smaller, whereas rocky platforms, being solid structures, are not listed. The understudied shore types lying in between these two extremes, can be categorised into very coarse sand (1 —< 2mm), granules (2 —< 4mm), pebbles (4 —< 64mm), cobbles (64 —< 256mm) and boulders (loose rock with a particle size of 256mm or larger).

Research on very coarse sand and granule shores are limited to Europe and largely comprise papers describing the behaviour of single taxa (de la Huz et al., 2002; Pubill et al., 2011; Fanini et al., 2019; Mathers et al., 2019), or ones that investigate physical characteristics, which can be broadly grouped into oceanography (Williams et al., 2002; Williams et al., 2003; Russell, 2006; Asano et al., 2010; Ruessink et al., 2015), or geology (Packman et al., 2001; Tõnisson et al., 2007; Bujan et al., 2019; Miroshnikova & Neradovsky, 2019; Nhon et al., 2019). Although some authors mention very coarse sand and granule shores as occurring in South Africa (Brown, 1971) no regional studies have focused solely on these shore types as habitats.

Cobble and pebble shore studies mostly originate in Europe (Packman et al., 2001; Bujan et al., 2019) or the United States of America (Matsumoto et al., 2019) in the field of geomorphology, while geological descriptions have been published in Russia (Miroshnikova & Neradovsky, 2019) and Estonia (Tõnisson et al., 2007). Single taxa studies have been undertaken in Japan (Kurihara, 2001, 2002, 2007; Nakakoa & Wada, 2017) and Brazil (Apolinário, 1999;

5

Furota & Ito, 1999), with a unique study on the relationship between algal diversity and disturbance conducted in Ghana (Liberman, 1979). Preliminary results from this study also helped the South African National Biodiversity Assessment to recognize pebble and cobble as being unique shore types (Griffiths & Robbins, 2019; Harris et al., 2019).

Boulder shores have been used to test ecological theories, as in California (Sousa, 1979a, b, 1980; McGuinness, 1984, 1987a, b, 1988) and Australia (Chapman, 2002, 2003, 2005, 2008), with several single taxa studies originating from Japan (Takada, 1992, 1995a, b, c, 1996, 2001, 2003, 2008) and South Africa (Henninger & Hodgson 2000; Henninger, 2001). Similar to pebble and cobble shores, only one study in South Africa (Tucker et al., 2017) reviews boulders as a habitat type, as such they have recently been included in the National Biodiversity Assessment (Sink et al., 2012).

The purpose of this chapter is to collate and compare former and current survey data from a range of intertidal shores in the Western Cape from across the full particle spectrum from fine sand to rocky shore types, and to objectively determine how many separate habitat types occur across this particle spectrum of shores and the degree of dissimilarity between each of these habitat types. Also, to identify which taxa, or groups of taxa, best characterise each of these identified habitat types.

Methods

Study Sites

This study collated data collected from intertidal transects undertaken across the entire spectrum

of sediment particle sizes from fine sand through to rocky shores. The geographical range of sites

included is confined to the region from Cape Agulhas to Jacobsbaai (Figure 2.1). Data for sandy

sites with grain sizes between 0.125mm and 1mm, as well as the majority of boulder and rocky

shores sites (>256+mm) were extracted from historical literature, while shores with particle grain

sizes 1 – 256mm were sampled during this study. As little to no literature exists on the

distribution of shingle or gravel shores (1 – 256mm) in South Africa, study sites were identified

using local knowledge accessed through consulting with local experts, social media,

communications with CapeNature and Sea Change, as well as by directly exploring the coastline.

A total of 58 sites were included in the study and are shown in Figure 2.1. GPS

coordinates, physical characteristics, data source and sampling dates of each site are listed in

Table 2.1. Due to graduated changes in median grain size occurring horizontally along the length

of several of the longer shores, more than one transect was sampled at Blouberg (five transects),

Grotto Bay and at Ganzekraal (both three transects). All transects, historical and sampled, were

treated as individual sites in the analyses.

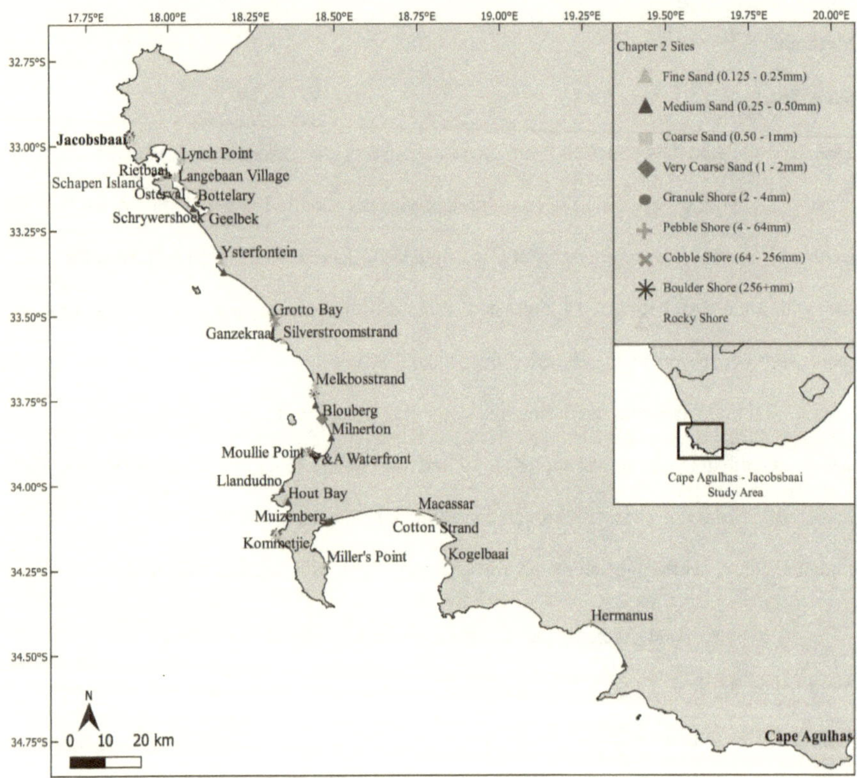

Figure 2.1 Map of the South West coast of South Africa showing the locations of the study sites (several transects were taken on shores with a range of grain sizes – see text and Table 2.1). Sites are marked with symbols indicating their shore types, as shown in the accompanying key. Created in QGIS v 3.10 (QGIS Development Team, 2019).

Table 2.1 The 58 sites, listed in order of increasing particle grain size, distinguished as either mean (Mn) or median (Md), with accompanying GPS coordinates, type of stratum, beach width (m), data source and sampling date of each study.

Site Name	Coordinates	Particle Size (mm)	Type of Stratum	Beach Width (m)	Source	Sample Date
Langebaan Village	33°05'06.1"S 18°01'46.4"E	0.12 Md	Fine Sand	79.2	Day, 1959	April 1946 - January 1958
Strand	34°06'16.5"S 18°48'55.7"E	0.14 Mn	Fine Sand	130.0	Nel, 2000	1995
Geelbek	33°11'34.0"S 18°07'25.8"E	0.14 Md	Fine Sand	400.0	Puttick, 1977	February 1974 - March 1975
Silverstroomstrand	33°34'17.2"S 18°20'40.4"E	0.16 Mn	Fine Sand	105.0	Soares, 2003	May 1992 - June 1993
Osterval (Langebaan Lagoon)	33°07'05.9"S 18°03'01.4"E	0.17 Md	Fine Sand	385.6	Day, 1959	April 1946 - January 1958
Harris Melkbosstrand	33°42'39.1"S 18°26'32.5"E	0.17 Mn	Fine Sand	148.0	Harris Unpublished	2010
Cotton	34°05'42.3"S 18°48'05.4"E	0.17 Mn	Fine Sand	86.0	Nel, 2000	1995
Macassar	34°04'44.3"S 18°45'03.2"E	0.17 Mn	Fine Sand	106.0	Nel, 2000	1995
Rietbaai	33°05'38.6"S 17°58'49.8"E	0.17 Md	Fine Sand	600.0	Puttick, 1977	February 1974 - March 1975
Hout Bay Harbour	34°02'49.2"S 18°20'55.9"E	0.18 Md	Fine Sand	55.0	Brown, 1971	September 1956
Bottelary	33°08'34.6"S 18°05'26.0"E	0.18 Md	Fine Sand	500.0	Puttick, 1977	February 1974 - March 1975
Bally Melkbosstrand	33°41'57.1"S 18°26'22.6"E	0.20 Mn	Fine Sand	90.0	Bally, 1983	August 1981
Ysterfontein	33°20'10.3"S 18°09'37.1"E	0.21 Mn	Fine Sand	59.7	Bally, 1983	August 1981
Yzerfontein Central	33°20'23.0"S 18°09'36.2"E	0.21 Mn	Fine Sand	110.0	Nel, 2000	1995
Hout Bay North	34°02'45.0"S 18°21'08.7"E	0.24 Md	Fine Sand	55.0	Brown, 1971	April 1956
Harris Muizenberg	34°06'26.5"S 18°28'25.7"E	0.25 Mn	Medium Sand	100.0	Harris Unpublished	2010
Yzerfontein North	33°19'13.3"S 18°09'19.2"E	0.28 Mn	Medium Sand	100.0	Nel, 2000	1995
Yzerfontein South	33°22'11.3"S 18°10'06.1"E	0.33 Mn	Medium Sand	88.0	Nel, 2000	1995
Hout Bay East	34°02'47.5"S 18°21'29.3"E	0.34 Md	Medium Sand	40.0	Brown, 1971	June 1956

Schrywershoek	33°10'53.8"S 18°04'31.0"E	0.34 Md	Medium Sand	350.0	Puttick, 1977	February 1974 - March 1975
Bloubergstrand	33°45'40.8"S 18°26'32.2"E	0.35 Mn	Medium Sand	39.0	Soares, 2003	May 1992 - June 1993
Nel Muizenberg	34°06'17.4"S 18°28'53.0"E	0.36 Mn	Medium Sand	81.0	Nel, 2000	1995
Harris Hermanus	34°31'36.2"S 19°22'28.2"E	0.40 Mn	Medium Sand	42.0	Harris Unpublished	2009
Milnerton	33°51'28.4"S 18°29'19.4"E	0.44 Md	Medium Sand	54.5	Brown, 1971	May - December 1956
Brown Muizenberg	34°06'06.6"S 18°29'20.2"E	0.44 Md	Medium Sand	36.0	Brown, 1971	April 1956
Llandudno	34°00'29.9"S 18°20'24.8"E	0.46 Md	Medium Sand	N/A	Brown, 1971	1956
Lynch Point 1	33°02'43.3"S 18°02'20.1"E	0.60 Md	Coarse Sand	32.0	Day, 1959	April 1946 - January 1958
Blouberg South 2	33°48'14.7"S 18°27'43.4"E	1.58 Mn	Very Coarse Sand	38.0	This Study	June 2020
Blouberg North 2	33°48'11.5"S 18°27'40.3"E	1.70 Mn	Very Coarse Sand	40.0	This Study	June 2020
Blouberg North 1	33°48'12.9"S 18°27'41.1"E	3.15 Mn	Granule	36.0	This Study	September 2019
Blouberg	33°48'12.9"S 18°27'41.3"E	5.58 Mn	Pebble	26.0	This Study	April 2019
Blouberg South 1	33°48'14.4"S 18°27'43.0"E	8.68 Mn	Pebble	30.0	This Study	September 2019
Ganzekraal South	33°31'46.0"S 18°19'02.5"E	29.62 Mn	Pebble	20.0	This Study	October 2019
Ganzekraal	33°31'22.5"S 18°19'09.4"E	34.23 Mn	Pebble	12.0	This Study	October 2020
Grotto Bay North	33°30'19.2"S 18°18'54.1"E	35.25 Mn	Pebble	15.0	This Study	September 2019
Grotto Bay	33°30'20.3"S 18°18'54.1"E	44.22 Mn	Pebble	19.0	This Study	September 2019
Hermanus	34°24'32.3"S 19°16'30.6"E	46.16 Mn	Pebble	23.0	This Study	June 2019
V&A Waterfront 1	33°54'00.2"S 18°25'32.4"E	83.25 Mn	Cobble	21.0	This Study	May 2019
Ganzekraal North	33°30'55.0"S 18°19'21.4"E	89.67 Mn	Cobble	17.0	This Study	September 2019
Miller's Point	34°14'02.4"S 18°28'28.8"E	93.75 Mn	Cobble	14.0	This Study	April 2019
Grotto Bay South	33°30'20.9"S 18°18'53.6"E	108.50 Mn	Cobble	60.0	This Study	August 2019
Kogelbaai	34°13'38.2"S 18°50'29.5"E	121.17 Mn	Cobble	17.2	This Study	May 2019

Moullie Point 1	33°53'56" S 18°24'41" E	247.33 Mn	Cobble	39.4	Tucker, 2012	March 2012
Tucker Melkbosstrand 1	33°43'41" S 18°26'18" E	275.58 Mn	Boulder	47.0	Tucker, 2012	April 2012
Kommetjie North 1	34°08'29" S 18°19'17" E	316.25 Mn	Boulder	54.5	Tucker, 2012	February 2012
V&A Waterfront 2	33°54'03.5"S 18°25'27.3"E	337.58 Mn	Boulder	15.0	This Study	May 2019
Kommetjie South 1	34°08'41" S 18°19'08" E	442.00 Mn	Boulder	40.0	Tucker, 2012	February 2012
Jacobsbaai South 1	32°58'27" S 17°53'00" E	491.33 Mn	Boulder	18.0	Tucker, 2012	August 2012
Jacobsbaai North 1	32°58'07" S 17°53'05" E	547.83 Mn	Boulder	17.0	Tucker, 2012	April 2012
Moullie Point 2	33°53'56" S 18°24'41" E	N/A	Rocky	39.4	Tucker, 2012	March 2012
Tucker Melkbosstrand 2	33°43'41" S 18°26'18" E	N/A	Rocky	47.0	Tucker, 2012	April 2012
Kommetjie North 2	34°08'29" S 18°19'17" E	N/A	Rocky	54.5	Tucker, 2012	February 2012
Kommetjie South 2	34°08'41" S 18°19'08" E	N/A	Rocky	40.0	Tucker, 2012	February 2012
Jacobsbaai South 2	32°58'27" S 17°53'00" E	N/A	Rocky	18.0	Tucker, 2012	August 2012
Jacobsbaai North 2	32°58'07" S 17°53'05" E	N/A	Rocky	17.0	Tucker, 2012	April 2012
Lynch Point 2	33°02'38.5"S 18°02'11.2"E	N/A	Rocky	17.5	Day, 1959	April 1946 - January 1958
Exposed Shapen Is.	33°05'20.6"S 18°01'11.8"E	N/A	Rocky	20.0 (estimated)	Day, 1959	April 1946 - January 1958
Sheltered Shapen Is.	33°05'35.8"S 18°01'17.0"E	N/A	Rocky	20.4	Day, 1959	April 1946 - January 1958

Sampling Protocol

For sites sampled during this study, shore width at low water of spring tide was measured to the nearest meter using a tape measure. Replicating sandy shore sampling efforts by Day (1959) and McLachlan (1996), a single transect was then conducted across the shore from the high water mark (most recent drift line) to the low water mark. Each transect consisted of eight equally-spaced 20 x 20cm quadrat samples, the distances between quadrats thus varied among transects, as shore widths varied among sites. Further replicating Day (1959) and McLachlan (1996), quadrats were excavated to spade depth (20cm). Deeper excavations could have resulted in the capture of more species, and almost certainly more individuals (as indeed is the case on sandy shores), but it was felt that sample depth should be consistent with that used when sampling other substrata. Larger particles (64 – 256mm) were individually washed in a bucket of seawater and macroscopic organisms removed. Smaller grains (4 – 64 mm) were placed into a bucket filled with seawater (that had been previously sieved through a net with 1mm mesh so as not to introduce extraneous organisms) swirled into suspension and then the water rapidly poured through a net with 1mm mesh. This was repeated three times, after which any remaining sediment was visually examined for larger or heavier organisms that might not have been elutriated. All organisms from each sample were placed in a labeled plastic bag before being transported to and frozen at the University of Cape Town and frozen pending further examination. To calculate mean grain size of very coarse sand, granule, pebble, cobble or boulder shores, Tucker et al. (2017) sampling efforts were replicated as 20 stones each from high, mid and low shore were randomly sampled on each transect and their length and width diameters were measured to the nearest millimeter and averaged to determine mean grain size per stone. A mean particle size per transect was then calculated from all 60 measurements combined. Recorded grain size for earlier studies were extracted from those sources and were obtained by a variety of different sieving techniques, as described in the source documents as listed in Table 2.1.

At the University of Cape Town, specimens from each sample were sorted into morpho-species and then identified to the closest taxonomic level possible using a WILD M5 dissecting stereo-microscope (Heerenbrugg, Switzerland) and appropriate local field guides as listed in Branch et al. (2010). All species were listed according to the names accepted in the World Register of Marine Species (WoRMS) (WoRMS Editorial Board, 2020). Species were counted for the number of individuals, drained on paper towel for 10 seconds and weighed wet on a MFD HL-100 balance (0 - 300g) to the nearest 0.01g. Specimens that could not immediately be identified to species level were preserved and kept in 70% ethanol and referred to taxonomic experts for identification where possible.

Previous Work Data Extraction

Previous work in the form of publications and student theses that conducted similar transect surveys across sites in the region were examined and physical and biological data extracted from these where possible. Physical data extracted were date of study, grain size, substratum type and shore width. Where grain size was given in phi it was converted to millimeter (Blott & Pye, 2001). Some authors reported a mean particle size per site, while others gave a median particle size, as noted in Table 2.1. Shore width was either given or was extracted from Figures showing transect profiles, with any shore widths given in feet converted to meters. Substratum type was either given or determined by mean or median grain size per site using the Wentworth scale (1922). Biological data included species presence, and, if given, wet biomass and abundance.

Some authors did not provide biomass and/or abundance in their studies while other authors

provided non-numerical data for abundance using descriptive terms such as 'common' or 'fairly

abundant' to describe taxa presence. Studies including rocky shore sites sometimes reported

biomass for certain taxa, such as sponges or reed worms, but gave abundance as percent

coverage of a quadrat. Authors provided biomass either as wet weight, dry weight, ash free dry

weight or acidified weight. The conversions from Field et al. (1980) and Riccardi and Bourget

(1998) were used whenever possible to uniformly list all biomass as wet weight. All species

names were listed under the names accepted in the World Register of Marine Species (WoRMS)

(Worms Editorial Board, 2020), so may not conform to the older names given in the original

publications.

Data Treatment

Study Sites

All site locations were plotted, mapped and presented using QGIS version 3.10 (QGIS

Development Team, 2019). It should be noted that sandy (0.125 –< 1mm), rocky and

intermediate particle grain size (1 –< 256mm) shore types with varying wave action where each

sampled on the West Coast, around the Cape Peninsula and throughout False Bay in order to

eliminate geography as a variable.

Biological Characteristics

Raw biomass (g) and raw abundance (number) for each taxon were converted into biomass as

grams per square meter and abundance to number of individuals per square meter. For data

collected in this study, which include sites with a mean particle grain size falling between 1 –

256mm, quadrats were treated as replicates and a conversion factor of x25 was used to convert numbers and biomass per sample to per square meter, based on the 20 x 20cm quadrat size used. Data from previously sampled sites, where necessary, were converted from numbers or biomass per linear meter to per square meter by dividing by shore width. Conversions to per meter squared were chosen over per linear meter as this study followed Tucker et al.'s (2017) sampling efforts and data analysis to compare shore types, and because measures were linear meter and are never used by rocky shore researchers. The total species list combining previous and sampled data included over 500 taxa (Appendix A and B). This list was reduced prior to analysis by removing rare species, these being defined as those recorded less than three times across all shore types. Such species may represent vagrant, scarce, or misidentified taxa. They are in any event of little value in similarity analyses, since they are consistently reported as zeroes across almost all sites (making these sites appear similar) and contribute little power to the analysis from a statistical perspective. Partial identifications (those made to only genus, family, class/phylum level, or to group common name e.g. fly larva) were also mostly discarded, but were retained if no other representative of that group was already identified in the database.

From the reduced species list, species presence was tabulated across sites of increasing grain size (mean or median mm) to identify species ranges across various shore types. Species richness (number), biomass (g/m^2) and density (#/m^2), where possible, were calculated per shore type. Total species richness (number), biomass (g/m^2) and density (#/m^2) per shore type were then plotted as bar plots in R (R Core Team, 2013; package *ggplot2* Wickham, 2016; package *gridExtra* Auguie, 2017; package *forcats* Wickham, 2020; package *svglite* Wickham et al., 2020; package *ggtext* Wilke, 2020).

Statistical Analysis

Primer v6 statistical software (Clark & Gorley, 2006) was used to produce a Bray Curtis similarity resemblance matrix based on species richness. Species richness was chosen as it is uniformly available for all sites, while not all authors provided biomass and abundance per site, and grain size was reported as either mean or median particle sizes, which cannot be converted one to the other. Cluster dendrograms and non-metric multidimensional scaling (nMDS) plots based on the Bray-Curtis similarity resemblance matrix were created to identify how the taxa grouped according to species richness and shore type, as well as to determine how similar these shore type communities were to one another. To address the significance of shore type similarity a non-parametric Analysis of Variance (ANOVA) Kruskal-Wallis test followed by the post hoc test (Clarke, 1993) were both performed in R (R Core Team, 2013; package *dunn.test* Dinno, 2017; package *gplots* Warnes et al., 2020) on species richness (#) and density (#/m²) only. Due to insufficient data availability for certain shore types, either due to low sample sizes within the study area or lack of quantitative data provided by previous studies, only species richness and density from six of nine shore types were included with coarse sand (0.5 –< 1mm), very coarse sand (1 –< 2mm and granule (2 –< 4mm) shore type data removed. No non-parametric ANOVA was performed on biomass (g/m²) as only five of nine shore types had sufficient data available. To identify the suite of species most characteristic of each shore type biota a Similarity of Percentage (SIMPER) analyses (Clarke, 1993) was conducted in Primer v6 (Clarke & Gorley, 2006). The top five characteristic taxa and their contributing percentages (%) per group emerging from the non-metric multidimensional scaling (nMDS) plots were used to better define the biota characteristic of each habitat type.

Results

Species Richness

When mean species richness (number) is plotted against particle size a 'u' shaped trend was observed (Figure 2.2), beginning with a moderately high species count circa #25 among fine to coarse sand shores, then dipping to <10 species present in shores with intermediate particle sizes (1 – 64mm), before increasing again towards the larger particle grain sizes of cobble, boulder and rocky shores, which have by far has the greatest mean species richness count at >50. Granule, the central most shore type, had the lowest mean species richness count of <5.

Biomass

Mean wet biomass (g/m²) was comparatively very low across all shores with smaller particle sizes (Figure 2.2), from fine sand through to pebble shores, all having biomass <100g/m². In the larger particle sizes there was a dramatic and progressive increase from about 1,000g/m² on cobble shores to over >2,000g m² on boulders with rocky shores having by far the highest biomass at >3,000g/m². Granule shores had the lowest biomass at <1g/m² (Figure 2.2).

Density

Following trends for mean species richness (number) and mean biomass (g/m²), mean density (#/m²) was highest on the larger particle grain shores from cobble to boulder, between 2,000 – 4,000#/m², while, fine sand through pebble shores all showed abundances <1,000#/m² (Figure 2.2). Of the smaller particle grain size shores, only fine sand, very coarse sand and pebble reached an abundance similar to the larger particle grain size shores at ~1,000#/m² (Figure 2.2).

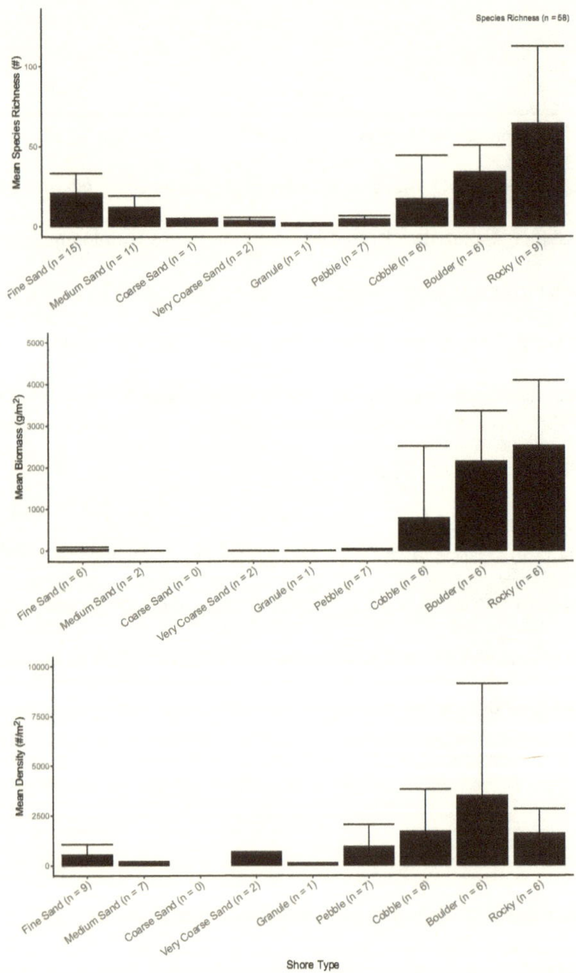

Figure 2.2 Bar plot with standard deviation error bars of mean species richness (#), mean biomass (g/m²), and mean density (#/m²) per shore type with n = number of sites. Coarse sand had a single site, Lynch Point 1, and the author, Day (1959), did not provide biomass or numerical abundance for his sites.

Shore Type Similarity

In the 2-Dimensional nMDS ordination plot shown in Figure 2.3, three major groupings occur

representing sandy shores, intermediate shores and larger particle grain size shores respectively.

Fine sand (0.125 –< 0.25mm) shore types split amongst two groupings, one with medium sand

(0.25 –< 0.50mm) shore types, and the other with a miscellaneous group of intermediate particle

sizes including one very coarse sand (1 –< 2mm), one pebble (4 –< 64mm), two cobble (64 –<

256mm) and one boulder (256+mm) shore types. The third grouping is cobble (64 –< 256mm),

boulder (256+mm) and rocky shore types (Figure 2.3). Two smaller groupings, one of rocky and

a single boulder shore and one of rocky shores also occur. Coarse sand (0.50 –< 1mm), granule

(1 –< 2mm) and most pebble shore types remain scattered (Figure 2.3). The V & A Waterfront 1

site is not shown on the plot as no fauna were present at this site, making it only very distantly

related to all other sites.

These groupings are similarly reflected in the cluster dendrogram (Figure 2.4) that shows three

major groupings occurring amongst the 58 sites at the 10% similarity level. Fine sand, medium

sand and coarse sand shores together form the first group, with very coarse sand, granule and

pebble shores together forming the second group, and boulder and rocky shores together forming

the third grouping. Cobble shores either group with the very coarse sand through pebble shore

group, or with the rocky and boulder shore group with one cobble site, V & A Waterfront 1 at

0% similarity with all other sites due to its lack of any fauna (Figure 2.4).

Both species richness (#) (K = 27.5345, df = 5, p = 4.487491e-05) and density (#/m²) (K =

20.4674, df = 5, p = 0.001020854) significantly differed among certain shore types (Table 2.2).

The Kruskal-Wallis post-hoc test revealed that of 15 shore type comparisons, only five shore type comparisons based on species richness (#) and seven shore type comparisons based on density (#/m²) were similar (i.e., p>0.05, Table 2.2). Among these only the shore type comparisons of boulder and rocky shores (species richness p value 0.22; density p value 0.452), cobble and pebble shores (species richness p value 0.137; density p value 0.454) and medium sand and fine sand shores (species richness p value 0.058; density p value 0.249) accepted the null hypothesis with p values greater than 0.05 for both post hoc test results, species richness (#) and density (#/m²).

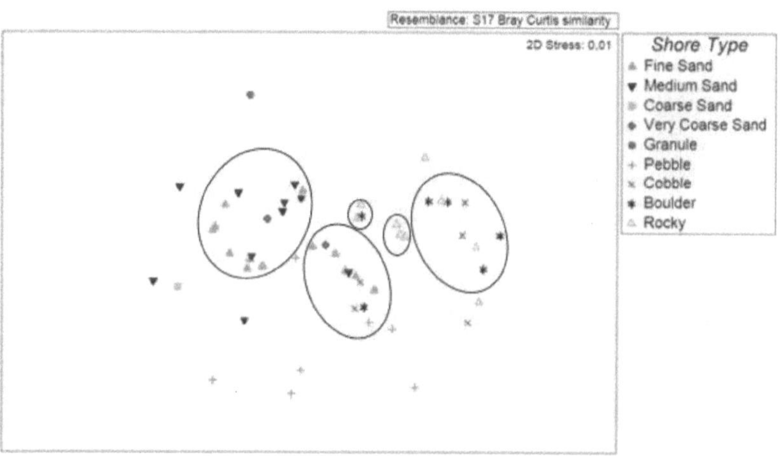

Figure 2.3. 2-Dimensional nMDS ordination plot (stress 0.01) of species richness (#) on 58 shore sites, which are colour coded according to shore type.

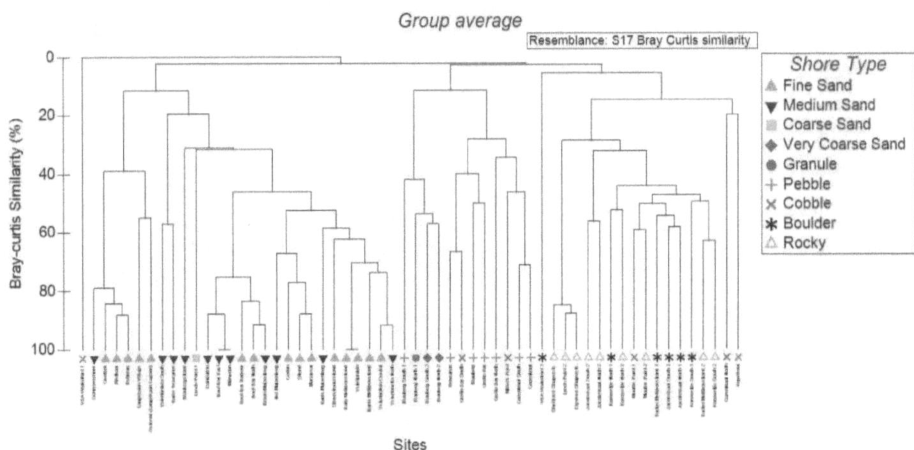

Figure 2.4. Cluster dendrogram of species richness (#) on 58 shore sites based on shore type showing how these group into three major clusters at the 10% similarity level

Table 2.2 Non-parametric Analysis of Variance (ANOVA) Kruskal-Wallis and post hoc test results for A: Species Richness (#) (K = 27.5345, df = 5, p = 4.487491e-05) and B: Density (#/m²) (K = 20.4674, df = 5, p = 0.001020854). Biomass (g/m²) was not tested due to insufficient data available for four of nine shore types. For both analyses H0 is that species richness (#) (or density (#/m²)) is the same among all shore types, and H1 is that species richness (#) (or density (#/m²)) differ among shore types.

A

	Comparisons	H0	H1	Z	p.value	Conclusion
15	Pebble - Rocky	Pebble = Rocky	Pebble =/ Rocky	-4.5	< 0.001	Reject H0
11	Medium Sand - Rocky	Medium Sand = Rocky	Medium Sand =/ Rocky	-3.25	0.001	Reject H0
13	Cobble - Rocky	Cobble = Rocky	Cobble =/ Rocky	-3.15	0.001	Reject H0
14	Fine Sand - Rocky	Fine Sand = Rocky	Fine Sand =/ Rocky	-1.99	0.024	Reject H0
12	Boulder - Rocky	Boulder = Rocky	Boulder =/ Rocky	-0.77	0.22	Accept H0
8	Boulder - Pebble	Boulder = Pebble	Boulder =/ Pebble	3.34	< 0.001	Reject H0
10	Fine Sand - Pebble	Fine Sand = Pebble	Fine Sand =/ Pebble	3.12	0.001	Reject H0
7	Medium Sand - Pebble	Medium Sand = Pebble	Medium Sand =/ Pebble	1.67	0.048	Reject H0
9	Cobble - Pebble	Cobble = Pebble	Cobble =/ Pebble	1.09	0.137	Accept H0
6	Cobble - Fine Sand	Cobble = Fine Sand	Cobble =/ Fine Sand	-1.7	0.045	Reject H0
4	Medium Sand - Fine Sand	Medium Sand = Fine Sand	Medium Sand =/ Fine Sand	-1.57	0.058	Accept H0
5	Boulder - Fine Sand	Boulder = Fine Sand	Boulder =/ Fine Sand	0.89	0.186	Accept H0
3	Boulder - Cobble	Boulder = Cobble	Boulder =/ Cobble	2.17	0.015	Reject H0
2	Medium Sand - Cobble	Medium Sand = Cobble	Medium Sand =/ Cobble	0.39	0.349	Accept H0
1	Medium Sand - Boulder	Medium Sand = Boulder	Medium Sand =/ Boulder	-2.08	0.019	Reject H0

B

	Comparisons	H0	H1	Z	p.value	Conclusion
14	Medium Sand - Rocky	Medium Sand = Rocky	Medium Sand =/ Rocky	-3.32	< 0.001	Reject H0
13	Fine Sand - Rocky	Fine Sand = Rocky	Fine Sand =/ Rocky	-2.85	0.002	Reject H0
15	Pebble - Rocky	Pebble = Rocky	Pebble =/ Rocky	-0.91	0.18	Accept H0
12	Cobble - Rocky	Cobble = Rocky	Cobble =/ Rocky	-0.77	0.22	Accept H0
11	Boulder - Rocky	Boulder = Rocky	Boulder =/ Rocky	-0.12	0.452	Accept H0
10	Medium Sand - Pebble	Medium Sand = Pebble	Medium Sand =/ Pebble	-2.5	0.006	Reject H0
9	Fine Sand - Pebble	Fine Sand = Pebble	Fine Sand =/ Pebble	-1.97	0.024	Reject H0
7	Boulder - Pebble	Boulder = Pebble	Boulder =/ Pebble	0.79	0.215	Accept H0
8	Cobble - Pebble	Cobble = Pebble	Cobble =/ Pebble	0.11	0.454	Accept H0
4	Boulder - Medium Sand	Boulder = Medium Sand	Boulder =/ Medium Sand	3.19	0.001	Reject H0
5	Cobble - Medium Sand	Cobble = Medium Sand	Cobble =/ Medium Sand	2.52	0.006	Reject H0
6	Fine Sand - Medium Sand	Fine Sand = Medium Sand	Fine Sand =/ Medium Sand	0.68	0.249	Accept H0
2	Boulder - Fine Sand	Boulder = Fine Sand	Boulder =/ Fine Sand	2.72	0.003	Reject H0
3	Cobble - Fine Sand	Cobble = Fine Sand	Cobble =/ Fine Sand	2.01	0.022	Reject H0
1	Boulder - Cobble	Boulder = Cobble	Boulder =/ Cobble	0.65	0.258	Accept H0

Diversity and Composition of Biota across Shore Type

A full listing of all the species found at the study sites is shown in Table 2.3. Examination of these data reveal that rocky shores have the highest species richness, with 171 taxa including attached taxa such as Mollusca, Porifera, Ascidiacea, Cnidaria and Algae, as well as other taxa vulnerable to desiccation such as Bryozoa, Echinodermata and Pisces (Table 2.3). Distantly following rocky shores are boulder shores with a count of 93 taxa comprised of similar groups to those found on rocky shores. Fine sand has a count of 88 taxa, mostly mobile burrowers with Polychaeta, Isopoda, Amphipoda, Bivalvia, and Gastropoda occurring at every site similar to cobble shores at a count of 86 taxa mostly mobile Polychaeta, Isopoda, Amphipoda, and Gastropoda. Species richness descends on both ends of the particle grain size spectrum through the Missing Middle shores towards granule shores, which has the lowest count at two taxa. With the exception of boulder shores, the Missing Middle shores low taxa counts mostly include mobile air-breathing species such as various 'worms' and Arthropoda. Polychaeta, Isopoda, and Amphipoda are all present in eight of the shore types, excluding coarse sand for Polychaeta and granule for Isopoda and Amphipoda, with Bivalvia and Gastropoda present in six, excluding coarse sand and granule for both, pebble for Bivalvia and coarse sand for Gastropoda (Table 2.3).

Of the major groupings seen in the nMDS plot (Figure 2.3) and cluster dendrogram (Figure 2.4), a SIMPER analysis was conducted for fine sand and medium sand, pebble and cobble and rocky and boulder (Table 2.4). Coarse sand and granule, having one site throughout the study, could not be used to generate a SIMPER plot while very coarse sand, which only had two sites, had a limited species list of just two contributing taxa. The characteristic taxa found within each of these groups are shown in Table 2.4. Fine sand and medium sand were mostly defined by mobile taxa including the Amphipoda, *Africorchestia quadrispinosa,* the Gastropoda, *Bullia digitalis,* the nemertean, *Cerebratulus fuscus,* and the Isopoda, *Exocirolana latipes* and *Eurydice barnardi,* each of them have a contributing percent over 2.7% (Table 2.4a). Pebble and cobble shores were characterised by mobile arthropod species, *Amaurobioides africanus, Capeorchestia capensis,* and *Ligia dilatata,* each contributing ~7% while the remaining two mobile aquatic species, *Paramoera capensis* and *Procereodes sp.,* contributed ~6% (Table 2.4b). Boulder and rocky shores were mostly characterized by algae, with four of the top five contributing species being algae, namely *Pachymenia orbitosa, Splachnidium rugosum, Champia lumbricalis* and *Sarcothalia stiriata,* each contributing <2% (Table 2.4c). The remaining species was the Gastropoda, *Scutellastra cochlear,* contributing 1.38%.

Table 2.3 Species recorded at each study site with species richness totaled per site and shore type.

The table records **Species Presence** across **Sites**, grouped by **Shore Type with Particle Grain Size Range (mm)**. Shore types: Fine Sand (0.125–<0.25), Medium Sand (0.25–<0.50), Coarse Sand (0.50–<1.00), Very Coarse Sand (1.00–<2.00), Granule (2.00–<4.00), Pebble (4.00–<64.00), Cobble (64.00–<256.00), Boulder (256.00+), Rocky (N/A). Presence is marked with •.

Fine Sand (0.125–<0.25)

Species/Taxa	Langebaan Village	Strand	Gordiek	Silwerstroomstrand	Ostewal (Langebaan Lagoon)	Harris Middeloosstrand	Cotton	Macassar	Bafant	Hout Bay Harbour	Robberg	Baly Middeloosstrand	Ysterfontein	Ysterfontein Central	Hout Bay North
Porifera															
Hymeniacedon perlevis															
Cnidaria															
Actinia ebhayiensis															
Aglaophenia pluma															
Anthothoe stimpsoni															
Bunodactis reynaudi															
Bunodosoma capense															
Corynactis annulata															
Isanthus capensis															
Kirchenpaueria pinnata															
Plumularia setacea															
Pseudactinia infecunda															
Platyhelminthes															
Notocomplana erythrotaenia															
Procerodes sp.															
Nemertea															
Cerebratulus fuscus	•		•	•		•	•	•	•	•	•			•	
Sipunculida															
Golfingia capensis															
Polychaeta															
Arenicola loveni	•			•					•						
Antinoe lactea	•			•											
Arabella tricolor															
Branchiomma violacea															

Medium Sand (0.25–<0.50) · Coarse Sand (0.50–<1.00) · Very Coarse Sand (1.00–<2.00) · Granule (2.00–<4.00)

Species/Taxa	Harris Melkbosberg	Ysterfontein North	Ysterfontein South	Hout Bay East	Schryvershoek	Melkbosstrand	Sol Melkbosberg	Harris Hermans	Milnerton	Brown Melkbosberg	Llandudno	Lynch Point 1	Blouberg South 2	Blouberg North 2
Porifera														
Hymeniacedon perlevis														
Cnidaria														
Actinia ebhayiensis														
Aglaophenia pluma														
Anthothoe stimpsoni														
Bunodactis reynaudi														
Bunodosoma capense														
Corynactis annulata														
Isanthus capensis														
Kirchenpaueria pinnata														
Plumularia setacea														
Pseudactinia infecunda														
Platyhelminthes														
Notocomplana erythrotaenia														
Procerodes sp.												•	•	•
Nemertea														
Cerebratulus fuscus	•	•												
Sipunculida														
Golfingia capensis														
Polychaeta														
Arenicola loveni				•										
Antinoe lactea														
Arabella tricolor														
Branchiomma violacea														

Pebble (4.00–<64.00) · Cobble (64.00–<256.00)

Species/Taxa	Blouberg North 1	Blouberg	Blouberg South 1	Groenkloof South	Groenkloof	Grotto Bay North	Grotto Bay	Romans	VAA Waterfront 1	Groenkloof North	Miller's Point	Grotto Bay South	Kogelbaai	Maselbe Point 1
Porifera														
Hymeniacedon perlevis														
Cnidaria														
Actinia ebhayiensis													•	•
Aglaophenia pluma														
Anthothoe stimpsoni														
Bunodactis reynaudi														
Bunodosoma capense											•			•
Corynactis annulata														
Isanthus capensis														•
Kirchenpaueria pinnata														
Plumularia setacea														
Pseudactinia infecunda														
Platyhelminthes														
Notocomplana erythrotaenia														•
Procerodes sp.	•													
Nemertea														
Cerebratulus fuscus									•					
Sipunculida														
Golfingia capensis														•
Polychaeta														
Arenicola loveni														
Antinoe lactea														
Arabella tricolor											•		•	
Branchiomma violacea														

Boulder (256.00+) · Rocky (N/A)

Species/Taxa	Tinker Middeloosstrand 1	Kommetjie South 1	VAA Waterfront 2	Kommetjie South 1	Jacobsbaai South 1	Jacobsbaai North 1	Maselbe Point 2	Tinker Middeloosstrand 2	Kommetjie North 2	Kommetjie South 2	Jacobsbaai South 2	Jacobsbaai North 2	Lynch Point 2	[illegible]
Porifera														
Hymeniacedon perlevis													•	•
Cnidaria														
Actinia ebhayiensis	•	•											•	•
Aglaophenia pluma													•	•
Anthothoe stimpsoni													•	•
Bunodactis reynaudi					•	•							•	•
Bunodosoma capense	•		•	•	•	•							•	•
Corynactis annulata													•	•
Isanthus capensis	•		•	•	•									
Kirchenpaueria pinnata													•	•
Plumularia setacea													•	•
Pseudactinia infecunda													•	•
Platyhelminthes														
Notocomplana erythrotaenia														
Procerodes sp.														
Nemertea														
Cerebratulus fuscus													•	
Sipunculida														
Golfingia capensis					•	•							•	•
Polychaeta														
Arenicola loveni														
Antinoe lactea					•								•	•
Arabella tricolor	•						•		•		•		•	•
Branchiomma violacea													•	•

...rmia sp.
...rmia ...lata
...magna
...ene sp.
...aphroditois
...sine capensis
...nas ...rei
...notus ...tus
...a tridactyla
...rea gaimardi
...tueris ...na
...tueris sp.
...e natalensis
...sa depressa
...is laevigata
...s capensis
...a hombergi
...a ...branchia
...ache ...alis
...astus ...na
...a ...quensis
...eospira ...rica
...adae, ...tified sp
...reis namibia
...huenopterus
...s arenosus
...ereis ...lii
...pio ...ata
...rrineris ...lerma
...nereis ...ta
...pis squamata
...ma tetraura
...os sp.
...n capensis

Species																								
Simplisetia erythraeensis	•	•		•	•			•																
Spirorbis spp.																		•	•		•		•	•
Steggoa capensis																					•			
Syllidae sp							•											•			•	•		
Telothelepus capensis	•	•	•	•			•																	
Thelepus pequenianus																								•
Arachnida																								
Amaurobioides africanus							•	•			•			•										
Desis formidabilis																•	•		•	•	•		•	•
Tanystylum brevipes																	•				•			•
Insecta																								
Anurida maritima								•	•			•	•											•
Dolichopodidae larvae		•		•	•																			
Fucellia capensis	•						•	•		•														
Telmatogeton minor															•		•			•				
Cirripedia																								
Amphibalanus amphitrite																								•
Austromegabalanus cylindricus																								
Balanus glandula																•	•		•	•	•	•		
Notomegabalanus algicola																								
Tetraclita serrata																•		•	•	•				•
Leptostraca																								
Nebalia capensis		•		•												•								
Tanaidacea																								
Zeuxoides helleri																•	•	•				•		
Isopoda																								
Cirolana venusticauda																•	•		•			•		•
Cymodocella sublevis																•				•				
Deto echinata								•		•	•			•		•								
Eurydice barnardi	•	•	•	• •		•																		
Eurydice kensleyi	•	•	• •	•			• •		•														
Eurydice longicornis	•		•		• •	•		• • •															
Eurydice sp.		•				•		•																
Excirolana latipes	•		• •	• •	•	• • • •	•	• • •															
Excirolana natalensis		•	•		• • •	•	• • •																
Exosphaeroma antikrausai															•		•	•						

...sphaeroma
coetes

...sphaeroma
...ssi

...sphaeroma
...fusculum

...sphaeroma
...um

...sphaeroma
...cattelzon

...sphaeroma
...color

...ropsis palpalis

...yromene huttoni

...yromene ovalis

...a dilatata

...atolana hirtipes

...idotea ungulata

...isocladus
...foratus

...isocladus
...psoni

...s grandatus

...phipoda

...corchestia
...drispinosa

...ryllis
...rophthalma

...pelisca palmata

...hyale
...dicornis

...yporeia gracilis

...liopiella
...haelseni

...eorchestia
...insis

...rella equilibra

...rella scaura

...adocus
...omaculatus

...nahusa filosa

...mopus rapax

...chestia
...senensis

...ianassa ceratina

...amoera capensis

...oculodes
...gimarus

...xocephalidae sp

...tohyale
...lanha

| Polycheria atolli |
| Ptilohyale plumulosa |
| Temnophlias capensis |
| Urothoë elegans |
| Urothoe grimaldii |
| Urothoë pinnata |
| **Mysidae** |
| Gastrosaccus psammodytes |
| Gastrosaccus sp. |
| **Decapoda** |
| Cryptodromiopsis spongiosa |
| Cyclograpsus punctatus |
| Danielella edwardsii |
| Diogenes brevirostris |
| Guinusia chabrus |
| Hymenosoma orbiculare |
| Kraussillichirus kraussi |
| Paguristes gamianus |
| Pilumnoides perlatus |
| Synalpheus tumidomanus |
| Upogebia africana |
| **Bryozoa** |
| Alcyonidium nodosum |
| **Bivalvia** |
| Aulacomya atra |
| Carditopsis rugosa |
| Choromytilus meridionalis |
| Donax serra |
| Mytilus galloprovincialis |
| Tellimya sp. |
| Tellimya trigona |
| Thecalia concamerata |
| Venerupis corrugata |
| **Polyplacophora** |

thochitona
oti

topleura
lo

ochiton bergoti

ochiton textilis

ta
virescens

oplax polita

ropoda

ittorina
saensis

buccinum
losum

a digitalis

a laevissima

a rhodostoma

apena
rhacta

apena cincta

apena
racea

apena spp.

ca dunkeri

nella sinuata

a mozambicus

ishla
ellana

ipatella
nsis

nula compressa

nula granatina

nula minuta

nula oculus

rafismrella
lhum

tropoma
linaceum

rella mutabilis

ula zonata

ton dunkeri

ton pectunculus

on prunona

arha
ssianus

arius speciosus

brilaria
hris

he rosea

Species
Myosotella myosotis
Nucella dubia
Nucella squamosa
Oxystele antoni
Oxystele tigrina
Scutellastra argenvillei
Scutellastra barbara
Scutellastra cochlear
Scutellastra granularis
Siphonaria capensis
Siphonaria compressa
Siphonaria concinna
Siphonaria serrata
Tricolia capensis
Tricolia neritina
Turbonilla kraussi
Turritella capensis
Thylacodes natalensis
Volvarina capensis

Cephalopoda

Octopus vulgaris

Echinodermata

Species
Amphipholis squamata
Amphiura capensis
Hemieonus insolens
Henricia ornata
Ophiothrix fragilis
Parechinus angulosus
Parvulastra exigua
Pentacta doliolum
Roweia frauenfeldi
Thyone aurea

Hemichordata

Balanoglossus capensis

Ascidiacea

Pyura stolonifera

Pisces

The table below is a single species-by-site presence matrix; it is shown here split into two column groups (left sites and right sites) with the taxon-label column repeated. Species-name labels are cropped at the left edge of the page and only their right-hand fragments are legible.

Left column group (sites 1–37)

Taxon																																					
…oclinus …a																																					
…ardia spp.																																					
…anthus …us																																					
…ceras …tum																																					
…um capense					•																																
…um planum					•																																
…ia …alis																																					
…hora …s																																					
…enia sinuosa																																					
…a maxima																																					
…na polycarpa																																					
…a spicifera																																					
…ria pallida																																					
…lla capensis																																					
…enia …a																																					
…enia …a																																					
…ra capensis																																					
…verrucosa																																					
…alia radula																																					
…alia striata																																					
…idum …n																																					
…es yendoi																																					
…mrum …forme																																					
…apensis																																					
Species Richness per Site	33	12	30	9	51	11	16	13	32	16	36	13	13	13	17	11	11	3	14	29	3	13	4	14	18	11	5	2	5	2	2	5	6	8	4	2	3
Species Richness per Shore Type	88															53											5	5		2	13						

Right column group (sites 38–58)

Taxon																					
…oclinus …a						•				•			•								•
…ardia spp.						•							•	•							
…anthus …us						•		•				•	•		•						
…ceras …tum						•				•											•
…um capense										•							•				
…um planum								•					•								
…ia …alis										•			•	•		•	•	•	•	•	•
…hora …s						•				•			•			•					•
…enia sinuosa						•							•								•
…a maxima																			•	•	•
…na polycarpa						•		•				•	•	•		•					•
…a spicifera						•						•									•
…ria pallida																			•	•	•
…lla capensis													•			•					•
…enia …a													•	•		•					•
…enia …a							•					•	•	•	•	•	•	•	•	•	•
…ra capensis							•	•					•	•	•	•	•	•	•	•	•
…verrucosa						•								•	•						
…alia radula																			•	•	•
…alia striata										•			•	•	•	•					•
…idum …n						•							•	•	•	•	•				•
…es yendoi						•	•	•		•			•	•	•	•		•			•
…mrum …forme												•	•								•
…apensis																			•	•	•
Species Richness per Site	0	8	8	3	12	72	45	33	2	45	33	46	63	41	28	35	15	17	127	121	130
Species Richness per Shore Type	86						93						171								

Table 2.4 SIMPER test results of mean dissimilarity of species richness (#) on **58** shore sites of varying shore types. Results shown reflect the top five contributing species of each MDS grouping, A: Fine Sand & Medium Sand, B: Pebble and Cobble, and C: Boulder and Rocky which both fall under 256+mm range as Rocky has no particle grain size. As such they are not compared, but rather grouped together. As both Coarse Sand and Granule shores only had one site, no SIMPER was generated, while Very Coarse Sand had a limited species list.

A

Groups Fine Sand and Medium Sand
Average dissimilarity = 66.09

Species	Group Fine Avg. Abund	Group Medium Avg. Abund	Avg. Diss	Diss/SD	Contrib. %	Cum. %
Africorchestia quadrispinosa	0.47	0.73	1.89	0.88	2.86	2.86
Bullia digitalis	0.53	0.55	1.84	0.82	2.79	5.65
Excirolana latipes	0.67	0.55	1.83	0.82	2.77	8.42
Cerebratulus fuscus	0.67	0.57	1.82	0.85	2.76	11.17
Eurydice barnardi	0.40	0.00	1.81	0.75	2.74	13.92

B

Groups Pebble and Cobble
Average dissimilarity = 83.59

Species	Group Pebble Avg. Abund	Group Cobble Avg. Abund	Avg. Diss	Diss/SD	Contrib. %	Cum. %
Amaurobioides africanus	0.43	0.17	6.08	0.53	7.27	7.27
Capeorchestia capensis	0.43	0.33	5.80	0.57	6.94	14.21
Ligia dilatata	0.57	0.50	5.77	0.58	6.90	21.11
Paramoera capensis	0.29	0.50	5.14	0.81	6.15	27.26
Procerodes sp.	0.57	0.00	4.95	0.80	5.92	33.18

C

Groups Boulder and Rocky
Average dissimilarity = 67.99

Species	Group Boulder Avg. Abund	Group Rocky Avg. Abund	Avg. Diss	Diss/SD	Contrib. %	Cum. %
Pachymenia orbitosa	0.00	0.89	1.24	1.12	1.82	1.82
Splachnidium rugosum	0.17	0.89	1.08	0.93	1.58	3.41
Champia lumbricalis	0.17	0.89	1.07	0.94	1.57	4.98
Scutellastra cochlear	0.17	0.67	0.94	0.78	1.38	6.36
Sarcothalia stiriata	0.17	0.56	0.92	0.76	1.36	7.72

Discussion

Species Richness, Biomass and Density

The aim of this study was to see how shores across a complete range of particle sizes differ in terms of their macrofaunal characteristics and how they group into broad habitat types. The results, most clearly seen in Figure 2.2, show that despite differences between individual site in terms of shore width, exposure and geographical location ect, similar trends occur across shore types for species richness (number), biomass (g/m²) and density (#/m²). All three parameters show by far the highest values occurring towards the larger particle size shores, pebble through rocky, and lowest values occurring on intermediate particle grain size shores, with granule shores having the lowest species richness (number), biomass (g/m²), and density (#/m²). These trends were expected, as granule shores are often reflective beaches characterized by steep slopes as a result of coarser particle grain sizes alongside low wave energy and microtidal range (Raffaeli & Hawkins, 1996; Short, 1996; McLachlan & Defeo, 2013; Bujan et al., 2019) creating difficult living conditions, with highly mobile particles that are too coarse to be easily burrowed into and too mobile to grip onto (Dexter, 1992). As such, coarser grain particle sizes are often less preferred by sandy shore inhabitants many of which are delicate, burrowing forms (de la Huz et al., 2002; Pubill et al., 2011; Fiora & Carcedo, 2015). Furthermore, boulder shores, which are often grouped with rocky shores due to a similar biotic composition, until recently compared by Tucker et al. (2017), offer a wide range of microhabitats such as between, beneath and on top of boulders (LeHir & Hily, 2005) that can be occupied by both attached (McGuinness 1987a, b) and mobile species, thus also allowing for higher species richness (number), biomass (g/m²) and density (#/m²) similar to rocky shores, as seen in Figure 2.2.

Shore Type Similarity

Three major groupings emerged amongst the nMDS plots results in Figure 2.3 and the cluster dendrogram in Figure 2.4. These are sandy (0.125 – 0.5mm) shores, multiple shore types falling within the Missing Middle particle grain size range (1 –< 256mm), and cobble, boulder and rocky (64 – 256+mm) shores, with two cobble shore sites included as outliers in the Missing Middle group. Additionally, the following three shore type groupings, fine sand and medium sand shore type groupings, pebble and cobble shore type groupings and boulder and rocky shore type groupings were not significantly different based on both the species richness (#) and density (#/m²) Kruskal-Wallis post hoc results (Table 2.2). Such groupings are not surprising as animals living across the nine shore types compared in this study tend to polarize at one of the two ends of the particle grain size spectrum, being either sandy shore burrowers or attached rocky shore inhabitants. Very few species can survive among the tumbling cobbles (Liberman, 1979) experienced on shores of 1 –< 256mm grain size, ultimately creating a group of shores with similar biotic composition but defined by just the few species that can survive here. Furthermore, it is likely these major groupings only formed at 10% similarity within the cluster dendrogram (Figure 2.4) given that this study considers the full spectrum of particle grain size across shore types with differences in biotic composition and grain size per shore type and as such the major groupings will only have similarity in lower percentages. Lastly, although scattered throughout the nMDS plots, pebble shores have not been grouped separately as nMDS plot can misrepresent groupings dependent on the perspective from which the multidimensional plots are viewed, hence cluster dendrograms are preformed alongside nMDS plots to further explore groupings (Van Sickle, 1997). When performed for this study, pebble shores grouped entirely with Missing Middle (1 –< 256mm) shores.

Biota Presence and Range across Shore Type

Similar to trends in mean species richness (number per site), biomass (g/m²) and density (#/m²)

(see Figure 2.2), the full tabulation of species presence at all sites combined in Table 2.3 shows

lowest numbers of species being found at shores of intermediate particle grain size (1 – 256mm)

with granule shores having the lowest species richness.

Taxa occurring across the full spectrum of shores are mostly comprised of worms and

Arthropoda, with Polychaeta occurring in all nine shore types and Amphipoda and Isopoda

occurring on eight shore types (although represented by different species in the different shore

types). A SIMPER analysis reveal species most characteristic of each shore type groups as being

mobile Isopoda and Amphipoda for fine sand, medium sand, pebble and cobble shores with

mostly attached algae characterising rocky and boulder shores.

The predominance of smaller burrowing animals on smaller grain particles is expected, given

that this habitat is soft and offers itself more readily to burrowing species, while the solid boulder

and rocky shores offer the stability required by attached species. Mobile animals, such as

Polychaeta, Amphipoda and Isopoda are seen across the entire spectrum of particle grain size

shores in this study, as both burrowing and mobile species occur within these taxa (Furota & Ito,

1999; LeHir & Hily, 2005; Weatherington, 2014; Pezy et al., 2016; Tucker et al., 2017).

However, shores with intermediate particle grain sizes (1 –< 256mm) have harsh conditions

(Raffaelli & Hawkins, 1996; Bujan, 2019) with rolling particle grain sizes (Liberman, 1979) that

do not allow for easy burrowing, mobility or attachment. Burrowing is easier in smaller grain particle sizes where burrowing rates are less limited by grain size (McLachlan, 1996; de la Huz et al., 2002; Pubill et al., 2011; Fiora & Carcedo, 2015), with medium sand being most preferred, as opposed to fine sand which is often smothering on the gills of burrowing animals (Wieser, 1959). The abundance of organic matter, rate of water flow and how aerobic or anaerobic interstices are also plays a role in species particle grain size preference (Snelgrove & Butman, 1994; Mehrshad et al., 2017).

Conclusion

Although Tucker et al. (2017) found boulder shores to be more biodiverse with higher species counts than rocky shores, the results of this study show the opposite. Possibly, this is due to the inclusion of the three rocky sites extracted from Day (1959), which were the subjects of a prolonged study conducted over several years and probably involving much more sampling effort than conducted by Tucker et al. (2017) who conducted a single transect at each site. This variation in methods and sampling efforts between the two studies allowed Day to collect taxa over all seasons and allows more opportunity for rare and vagrant species to be included in his surveys. Furthermore, two of Day's (1959) rocky sites, Exposed Schapen Island and Sheltered Schapen Island, were sheltered from wave action usually resulting in fewer species occupying these shores than exposed rocky shores as stated by Day (1959) and follows the Intermediate Disturbance Hypothesis (Connell, 1978). However, as these sites also occurred on a rocky shore which probably incorporated some boulders, it is possible that these shores were more biodiverse than other shores classified as rocky regardless of exposure as they harboured fauna from both rocky and boulder shore habitats. At the time of Day's (1959) study, boulder shores were not considered a separate shore type, and were simply included with rocky shores. As such, it is possible that boulder species were collected in these surveys and listed as rocky shore species. Furthermore, it could be argued that many of the taxa extracted from Day's (1959) sandy sites were actually more characteristic of rocky shores and should therefore, not be included in this study. Rocky shore species presence on Day's (1959) sandy sites is a result of *Pyura herdmani* and *Pyura stolonifera*, ascidians which embed and anchor themselves in the sandy shores at Langebaan Lagoon acting as hard substrates which allow for rocky shore taxa to occupy sandy shore habitats (Rius & Teske, 2011).

Previous to this study, the intermediate group of particle grain size shores (1 – 256mm) were unrecognized in South Africa, but this study shows them to be a well-defined and distinctly differing habitat type. Although these shore types are not rich in species they should be more fully investigated and this will be the topic of the next chapter, which explores the biota of two of the five intermediate shore types, pebble and cobble shores.

Chapter 3 - Breaking the rules:

Exploring the unusual biota of pebble and cobble shores

Introduction

The substratum of intertidal shores can be comprised of a full spectrum of particle sizes, ranging from fine muds and silts through to solid rocky platforms. Conventionally, shore types are categorized by Wentworth's (1922) particle size classification, according to which sediments < 2mm are classified as muds and sands, those 2 –< 4mm as granules, those 4 –< 64mm as pebbles, those 64 –< 256mm as cobbles and those > 256mm as boulders.

The vast majority of intertidal research has been undertaken on shores comprised either of fine sand/mud or of solid rock (Raffaelli and Hawkins, 1996), while limited research has been published on boulder shores (Sousa 1979a; McGuiness 1984, 1987a, b, and 1988; McGuinness & Underwood 1986; Chapman & Underwood, 1996; Barnes & Lehane, 2001; Chapman 2002, 2003, 2005, 2008, 2012, 2013 & 2017; Tucker et al. 2017) and even less on pebble and cobble shores (Apolinário, 1999; Furoata & Ito, 1999; Kurihara 2001, 2002, 2007; Ryu et al., 2012). Moreover, the few papers that do exist on pebble and cobble shore systems have focused on their geology (Packman et al, 2001; Tõnisson et al., 2007; Bujan et al., 2018; Bujan et al., 2019; Matsumoto, Young & Guza, 2019; Miroshnikova and Neradovsky, 2019), or on the vegetation along shingle or gravel shores, an interchangeable term for a collection of shore types from 1 –< 256mm (Gauci et al., 2005; Forgie et al., 2012), particularly those in the United Kingdom (Fuller, 1987; Davy et al., 2001; Packman & Soiers, 2001) and in New Zealand (Randall, 2008). Ecologically, a unique paper in Ghana correlates high diversity of algae growth on cobbles to disturbance as increased seasonal rain surge creates tumbling (Lieberman, 1979). Other studies examine the biology of single species that happen to occupy cobble habitats, for example a series of marine invertebrate-focused papers from Japan and Brazil (Apolinário, 1999; Furoata & Ito, 1999; Kurihara et al, 2001, 2002, 2007) and a study in China, exploring the pebble size preference of the decapod, *Cyclograpsus pumilio* (Nakakoa & Wada, 2017). Remarkably, of the few existing studies on pebble and cobble shores, only two examine these shores as ecosystems, by investigating the conservational needs of shingle shore invertebrates, mostly terrestrial Arthropoda (Shardlow, 2001) in the UK and a combination of terrestrial and marine invertebrates, mostly Polychaeta, Gastropoda, and Amphipoda in the Maltese islands (Gauci, 2005).

Pebble and cobble shores tend to be exposed to heavy wave action (Raffaelli and Hawkins, 1996) and have steeper slopes than most sandy or rocky shores (Bujan et al., 2019). The resultant particles are therefore mobile and rounded through erosion by wave action which causes grinding (Toanisson et al., 2007; Bujan et al., 2019; Miroshnikova and Neradovsky, 2019). Given these rough conditions and characteristic lack of visible macrofauna, pebble and cobble shores are very rarely surveyed for biota (Brown, 1971), as few species are thought to be able to survive such harsh environments. Unlike fine sandy shores, whose grain is small and light, these shore types have granules that are too large and heavy for many burrowing animals to find refuge within (Bally, 1981). Additionally, unlike larger boulders or stable rock platforms, pebble and cobble shores comprise constantly rolling particles too mobile for sedentary species to attach themselves to (McGuinness 1987a, b).

Within South Africa, no published literature exists on pebble or cobble shores, nor do their abundance or distribution patterns appear to have been surveyed. Therefore, the biota and conservational significance of such shores remains completely unknown. Indeed, the existence of these shore types has only recently been recognized in shore classification schemes, such as the

National Biodiversity Assessment (Griffiths & Robbins, 2019; Harris et al., 2019), due to the preliminary findings of this study.

This chapter will primarily focus on describing the biota found on pebble shores (grain size 4 –< 64mm) and cobble (grain size 64 –< 256mm) shores in the Western Cape region of South Africa. The specific aims are to quantify and describe the biota found on pebble and cobble shores in this region, to determine the intertidal distribution patterns of the component species, and to discuss their origins and adaptations to life in this habitat.

Methods

Study Sites

The study sites included shores of microtidal systems with grain size in the range 4 —< 256mm

found between Cape Agulhas and Jacobsbaai (a coastline of approximately 350km). As pebble or

cobble shore exists have not been previously mapped in South Africa, study sites were identified

using local knowledge accessed through personal contact with other researchers, social media,

communications with CapeNature and Sea Change, as well as by directly exploring the coastline.

A total of 12 transect sites were included in the study, distributed across seven beaches

(Figure 3.1). Due to graduated changes in median grain size occurring along the length of several

of the longer shores, more than one transect showing different grain sizes could be sampled at

Blouberg (two transects) and at Grotto Bay and Ganzekraal (both three transects). It should also

be noted that some sites had larger isolated boulders near the water line, as such some rocky and

boulder fauna may be present in low shore samples for certain sites, although this fauna is not

typical of pebble and cobble shores. Each of the transects was treated as a replicate in the

analyses. Sampling dates, GPS Coordinates and other physical characteristics of each site are

given in Table 3.1. Figure 3.2 illustrates a typical cobble shore (A) and a typical pebble shore

(B).

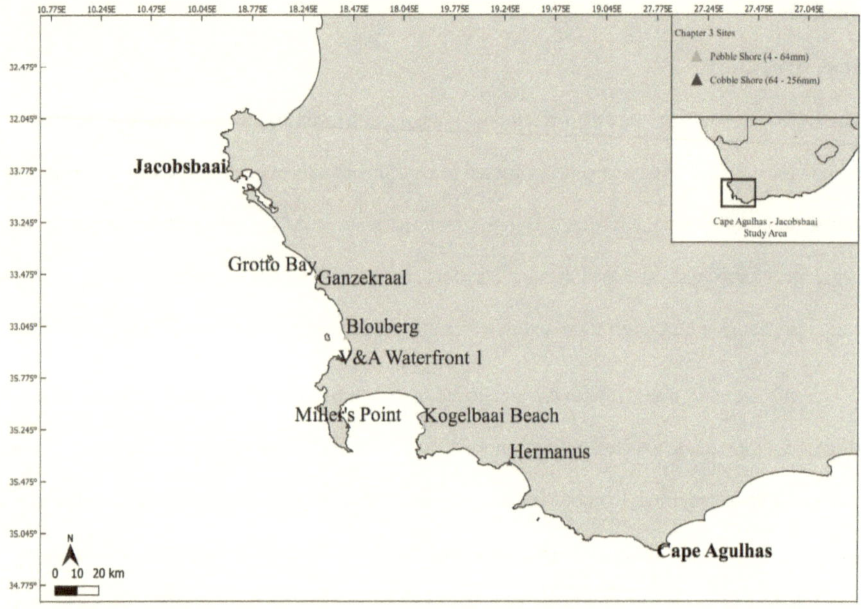

Figure 3.1 Map of the South West coast of South Africa showing the locations of the study sites. Created in QGIS v 3.10 (QGIS Development Team, 2019).

Table 3.1 The 12 sample sites, listed in order of increasing mean particle size (mm) with accompanying GPS coordinates, beach width (m) and sample date of each. The horizontal line separates pebble (4 –< 64mm, above) from cobble (65 –< 256mm, below) sites.

Site Name	Coordinates	Mean Particle Size (mm)	Beach Width (m)	Sample Date
Blouberg	33°48'12.9"S 18°27'41.3"E	5.58	26.0	7/4/2019
Blouberg South	33°48'14.4"S 18°27'43.0"E	8.68	30.0	13/9/2019
Ganzekraal South	33°31'46.0"S 18°19'02.5"E	29.62	20.0	15/10/2019
Ganzekraal	33°31'22.5"S 18°19'09.4"E	34.23	12.0	15/10/2019
Grotto Bay North	33°30'19.2"S 18°18'54.1"E	35.25	15.0	16/9/2019
Grotto Bay	33°30'20.3"S 18°18'54.1"E	44.22	19.0	16/9/2019
Hermanus	34°24'32.3"S 19°16'30.6"E	46.16	23.0	17/6/2019
V&A Waterfront 1	33°54'00.2"S 18°25'32.4"E	83.25	21.0	17/5/2019
Ganzekraal North	33°30'55.0"S 18°19'21.4"E	89.67	17.0	30/9/2019
Miller's Point	34°14'02.4"S 18°28'28.8"E	93.75	14.0	8/4/2019
Grotto Bay South	33°30'20.9"S 18°18'53.6"E	108.50	60.0	13/8/2019
Kogelbaai	34°13'38.2"S 18°50'29.5"E	121.17	17.2	21/5/2019

Figure 3.2 Photographs showing two of the sample sites. A: pebble shore at Kogel Bay (34°13'38.2"S 18°50'29.5"E), and B: cobble shore at Miller's Point (34°14'02.4"S 18°28'28.8"E).

Sampling Protocol

At each sampling site, shore width at low water of spring tide was measured to the nearest meter using a tape measure. Replicating sandy shore sampling efforts by Day (1959) and McLachlan (1996), a single transect was conducted across the shore from high water mark (most recent drift line) to low water mark, each transect consisted of eight equally-spaced 20 x 20cm quadrat samples. The distances between quadrats thus varied among transects, as shore widths varied among sites. Also, as shore profiles and tidal ranges differed somewhat between sites, quadrats of the same numerical sequence did not necessarily occur at the exact same tidal height (although these would be similar). Quadrats were excavated to a depth of 20cm, replicating the usual sample depths ('spade depth' or 20cm) used by most sandy shore researchers (Day, 1959; McLachlan, 1996). It should be noted that deeper excavations could include more species, and would certainly have include more individuals, however the top 10cm of depth includes the majority of fauna. It should also be noted that fauna living amongst these coarser particle grain sizes (1 – 256mm) tend to be very mobile and therefore difficult to sample. Larger grains (64 – 256mm) were individually washed in a bucket of seawater and macroscopic organisms removed. Smaller grains (4 – 64 mm) were placed into a bucket filled with seawater previously sieved through a net with 1mm mesh, swirled into suspension and then the water rapidly poured through a net with 1mm mesh This was repeated three times, after which the material in the bucket was also visually examined for larger and heavier organisms that might not have been elutriated. The organisms recovered from each sample were placed in a labeled plastic bag before being transported to the University of Cape Town and frozen for later analysis. To calculate mean grain size per transect 20 stones each from each of high, mid and low shore were randomly sampled and their length and width diameters measured to the nearest millimeter and averaged to determine mean diameter. The mean particle size per transect was then calculated as the average of all 60 measurements combined.

At the University of Cape Town specimens from each sample were sorted into morphospecies and identified to closest possible taxonomic level using appropriate regional field guides as listed in Branch et al. (2010) and a WILD M5 dissecting stereo-microscope (Heerenbrugg, Switzerland). All species were listed as names accepted in the World Register of Marine Species (WoRMS) (WoRMS Editorial Board, 2020). Species were counted, drained on paper towel for 10 seconds and weighed wet on a MFD HL-100 gram balance (0 – 300g) to the nearest 0.01g. Specimens that could not be identified to species level were preserved and kept in 70% ethanol and later referred to taxonomic experts for identification where possible.

Data Treatment

Study Sites

All site locations were plotted, mapped and presented using QGIS version 3.10 (QGIS Development Team, 2019).

Biological Characteristics

Raw biomass (g) and raw abundance (number) for each taxon was converted into biomass as grams per square meter and abundance as number of individuals per square meter (conversion factor x25). Species richness (number), biomass (g/m^2) and density ($\#/m^2$) were then calculated per tidal level per site. It should be noted that although quadrats did not share the same tidal height, they shared the same tidal level. This study followed Day's (1959) methods in that the ratio of distance between each of the eight quadrats along the transect line was equally spaced.

Therefore, pooling of species richness (#), biomass (g/m²) and density (#/m²) happened across tidal levels of varying tidal heights. Mean biomass (g/m²) and mean density (#/m²) for all species combined per tidal level were plotted as box and whiskers, while species richness (number) per tidal level was plotted as bar graphs in R (R Core Team, 2013; package *ggplot2* Wickham, 2016; package *gridExtra* Auguie, 2017; package *forcats* Wickham, 2020; package *svglite* Wickham et al., 2020; package *ggtext* Wilke, 2020).

To determine which were the most significant taxa characterizing these shores, an Index of Relative Importance (IRI) was calculated for each species using the formula IRI = f (n% + w%), where f = frequency of occurrence (percent of sites where that species was found), n = average percentage by numbers of that species, and w = average percentage by biomass of that species (Cunha & Cunha, 2016).

Statistical Analysis

Primer v6 statistical software (Clarke & Gorley, 2006) was used to plot accumulation curves, to ensure that adequate sampling occurred on both pebble and cobble shores. The separate pebble and cobble shore accumulation curve plots were combined into a single accumulation curve plot using Microsoft® Excel for Mac (version 16.39). The density (#/m²) data were fourth root transformed to account for differences among commonly and rarely occurring species prior to analysis (Clarke & Gorley, 2006). Primer v6 (Clarke & Gorley, 2006) was also used to produce a Bray-Curtis similarity resemblance matrix based on mean grain particle size per site and species abundance. Cluster dendrograms and non-metric multidimensional scaling (nMDS) plots based on the Bray-Curtis similarity resemblance matrix were created to determine how similar pebble and cobble shore communities were to one another, and whether or not these shore types should be treated as replicates.

Results

Sampling Effort

The curve for the pebble plot approached an asymptote, suggesting that the 56 samples taken across seven sites for pebble shores was a sufficient sampling effort to adequately assess the composition of the fauna, although not to sample all potential species present in these habitats, while the 40 samples taken across five sites for cobble shores could benefit from more sampling (Figure 3.3). Of the 96 samples taken across 12 sites, 29 species were collected from cobble shores while 22 were collected from pebble shores (Figure 3.3).

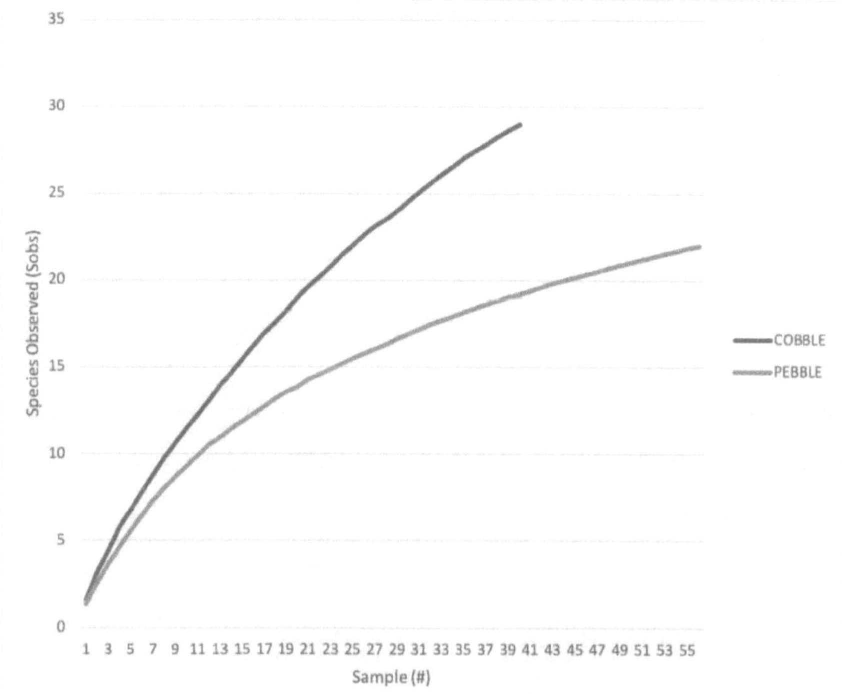

Figure 3.3 Accumulation Curves plotted using Species Observed (Sobs) for 56 samples taken from seven sites for pebble shores and 40 samples taken from five sites for cobble shores.

Shore Type Similarity

An nMDS plot showing the similarities between the seven pebble shores (4 – 64mm) and five

cobble shores (64 – 256mm) is shown in Figure 3.4. The V & A Waterfront 1 site is not shown

on the plot as no fauna were recovered from this site, making it only very distantly related to all

other sites. No groupings occurred amongst the sites, suggesting there is no distinction between

the faunal characteristics of the pebble and cobble shores sampled.

These groupings are further reflected in the form of a cluster dendrogram in Figure 3.5. The

dendrogram shows no separation between pebble and cobble shores, with the majority of sites

being between 10 - 40% similar to one another regardless of their shore type. The highest

similarities occurred between Hermanus and Grotto Bay South at 55%, and at 60% for

Ganzekraal and Ganzekraal South, both of which occurred on the same pebble shore. Only V &

A Waterfront 1 had a 0% similarity with all of the other sites due to the site having a lacking any

fauna.

Based on these results, it can be concluded that the pebble and cobble shores sampled comprise a

single habitat type and thus, all sites sampled across both particle size groupings are treated as

replicates in further analyses.

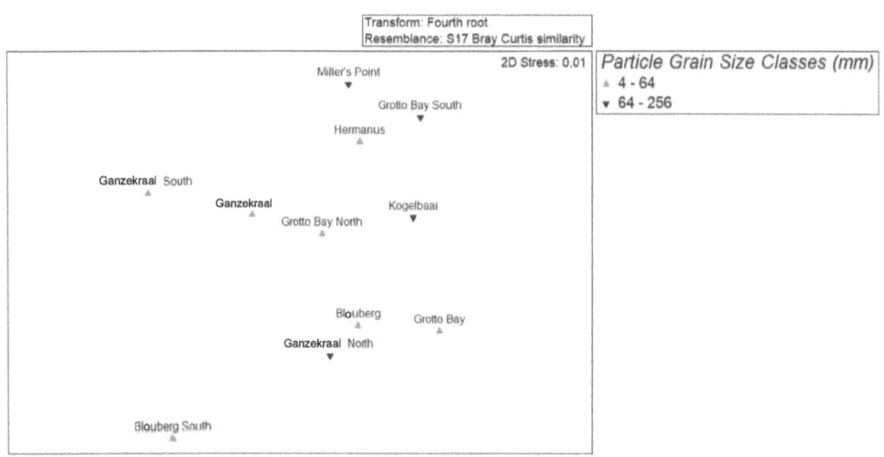

Figure 3.4 2-Dimensional nMDS ordination plot (stress 0.01) of fourth root transformed species density (#/m²) on 12 pebble (4 –< 64mm) and cobble (64 –< 256mm) shore sites based on particle grain size (mm) classes.

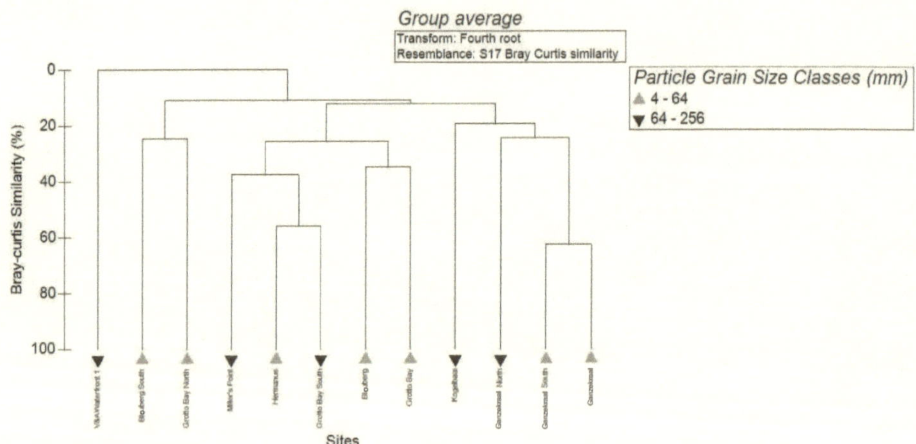

Figure 3.5 Cluster analysis dendrogram of fourth root transformed species density (#/m²) on 12 pebble (4 –< 64mm) and cobble (64 –

< 256mm) shore sites based on particle grain size (mm) classes.

Overall Composition of the Biota

A total of 39 faunal species were recorded in the surveys and are listed in Tables 3.2 and 3.3. The majority of these species fell within either Arthropoda (18 species; including six Insecta, five Amphipoda, five Isopoda, one Arachnida, and one Chilopoda) or Gastropoda (10 species). These groups together thus made up 72% of the fauna. The remaining species comprised four Polychaeta, two Nemertea, one Bivalvia, one Polyplacophora, one Platyhelminthes, one Echinodermata and one Anthozoa. Notably, macroalgae were completely absent from the survey, as were many other higher taxa typically associated with rocky or boulder shores (such as Porifera, Bryozoa, Ascidiacea, etc). Another notable aspect of the faunal mix was that 14 of the 39 species were semi-terrestrial air-breathing forms, mostly Arthropoda, including all Insecta, Arachnida and Chilopoda as well as four of five Isopoda, and one Amphipoda, and the pulmonate Gastropoda *Myosotella myosotis* (Table 3.2). The remaining taxa were aquatic in habitat. Also, there was only two sedentary forms, the Anthozoa, *Bundosoma capense*, and the Bivalvia, *Aulacomya atra* being the only such forms represented, while all other species recorded were mobile. Only 14 of the 39 species occurred more than once across all sites and are therefore considered characteristic of the fauna of pebble and cobble shores (Appendices C, D and E).

The degree to which each species contributed to the overall composition of the fauna of pebble and cobble shores can be calculated using an Index of Relative Importance or IRI (Cunha & Cunha, 2016) for each species across all 12 sites. These results show that the semi-terrestrial Isopoda, *L. dilatata,* was by far the most dominant member of the fauna, with by far the highest IRI value of 6078 (because it occurred in 58% of all sites and made up an average of > 60% of numbers and > 40% of biomass at each site). This was distantly followed by the pulmonate

Gastropoda, *M. myosotis,* with an IRI of 620, the semi-terrestrial Amphipoda, *T. capensis,* at

280, another semi-terrestrial drift-line Isopoda, *Deto echinata,* at 163, then the Gastropoda,

Helcion pruinosus, at 105, and Isopoda, *Marioniscus spatulifrons,* at 104. It is notable that

almost all of these were air-breathing forms, with *H. pruinosus* being the only aquatic species

with an IRI values over 100 (Table 3.2).

Table 3.2. Listing of the species found on pebble and cobble shores showing percentage occurrence (proportion of sites where found), average percentage contribution to biomass with standard deviation and average percentage contribution to density with standard deviation, as well as IRI of each species occurring across all 12 sites surveyed. Species listed in alphabetically under groups.

Index of Relative Importance (%) per Species Across Sites IRI = f % (n% + w%)							
Species Accumulation				*Total Occurrence*	*Average Density*	*Average Biomass*	IRI
Group	*Scientific Name*	*Common Name*	*Origin*	%	% [+/- SD]	% [+/- SD]	%
Anthozoa	*Bunodosoma capense*	Sea Anemone	Aquatic	8.33	0.03 [+/-180.42]	0.76 [+/-214.70]	6.61
Platyhelminthes	*Procerodes sp.*	Flatworm	Aquatic	33.33	1.42 [+/-5969.93]	0.04 [+/-7.74]	48.68
Nemertea	*Cerebratulus fuscus*	Nemertean	Aquatic	8.33	0.03 [+/-180.42]	0.04 [+/-10.83]	0.56
	Nemertean unidentified	Nemertean	Aquatic	8.33	0.09 [+/-541.27]	0.06 [+/-18.04]	1.26
Polychaeta	*Arabella iricolor*	Polychaete	Aquatic	8.33	0.06 [+/-360.84]	1.04 [+/-292.28]	9.15
	Perinereis namibia	Polychaete	Aquatic	33.33	2.26 [+/-7308.29]	0.72 [+/-149.11]	99.58
	Pseudonereis variegata	Polychaete	Aquatic	8.33	0.03 [+/-180.42]	0.30 [+/-82.99]	2.70
	Syllidae unidentified	Polychaete	Aquatic	8.33	0.06 [+/-360.84]	0.01 [+/-1.80]	0.54
Arachnida	*Amaurobioides africana*	Spider	Terrestrial	33.33	0.29 [+/-954.70]	0.04 [+/-5.63]	11.13
Chilopoda	*Geophilomorpha sp.*	Centipede	Terrestrial	16.67	0.09 [+/-388.49]	0.01 [+/-2.43]	1.66
Insecta	*Anurida maritima*	Collembolan	Terrestrial	16.67	0.90 [+/-5589.31]	0.02 [+/-5.40]	15.38
	Cercyon maritimus	Beetle	Terrestrial	8.33	0.03 [+/-180.42]	0.01 [+/-1.80]	0.29
	Fucellia capensis	Fly	Terrestrial	8.33	0.12 [+/-721.69]	0.02 [+/-5.41]	1.12
	Stratiomyidae (larvae)	Fly Larvae	Terrestrial	8.33	0.20 [+/-1080.16]	0.03 [+/-7.28]	1.95
	Staphylinidae unidentified	Beetle	Terrestrial	16.67	0.06 [+/-360.84]	0.01 [+/-1.80]	1.07
	Tabanidae (larvae)	Fly Larvae	Terrestrial	8.33	0.03 [+/-180.42]	0.22 [+/-63.15]	2.11
Isopoda	*Deto echinata*	Isopod	Terrestrial	25.00	1.66 [+/-10083.96]	4.88 [+/-1361.26]	163.49

	Species	Common name	Habitat				
	Exosphaeroma truncatitelson	Isopod	Aquatic	8.33	0.14 [+/-902.11]	0.01 [+/-3.61]	1.31
	Ligia dilatata	Isopod	Terrestrial	58.33	63.47 [+/-311496.81]	40.72 [+/-8251.66]	6077.84
	Marioniscus spatulifrons	Isopod	Terrestrial	8.33	4.16 [+/-25980.76]	8.30 [+/-2332.86]	103.84
	Philoscia sp.	Isopod	Terrestrial	8.33	0.17 [+/-1082.53]	0.01 [+/-1.80]	1.50
Amphipoda	*Eorchestia dassenensis*	Amphipod	Aquatic	8.33	0.88 [+/-4474.71]	0.17 [+/-37.70]	8.81
	Paramoera bidentata	Amphipod	Aquatic	41.67	0.32 [+/-1984.64]	0.04 [+/-10.83]	14.85
	Paramoera capensis	Amphipod	Aquatic	25.00	0.39 [+/-1195.63]	0.05 [+/-5.49]	10.80
	Ptilohyale plumulosa	Amphipod	Aquatic	25.00	0.20 [+/-728.66]	0.04 [+/-6.17]	5.97
	Talorchestia capensis	Amphipod	Terrestrial	41.67	5.06 [+/-20172.30]	1.66 [+/-273.74]	280.00
Bivalvia	*Aulacomya atra*	Bivalve	Aquatic	8.33	0.03 [+/-180.42]	2.77 [+/-777.62]	23.29
Polyplacophora	*Acanthochitona garnoti*	Chiton	Aquatic	8.33	0.06 [+/-360.84]	0.89 [+/-250.79]	7.92
Gastropoda	*Burnupena cincta*	Whelk	Aquatic	8.33	0.03 [+/-180.42]	3.91 [+/-1098.77]	32.81
	Burnupena lagenaria	Whelk	Aquatic	8.33	0.03 [+/-180.42]	2.94 [+/-826.33]	24.74
	Cymbula miniata	Limpet	Aquatic	8.33	0.03 [+/-180.42]	5.11 [+/-1436.16]	42.81
	Cymbula oculus	Limpet	Aquatic	8.33	0.03 [+/-180.42]	0.37 [+/-104.64]	3.34
	Helcion pectunculus	Limpet	Aquatic	8.33	0.03 [+/-180.42]	0.07 [+/-19.85]	0.83
	Helcion pruinosus	Limpet	Aquatic	16.67	0.69 [+/-3778.30]	5.60 [+/-1485.04]	104.84
	Myosotella myosotis	Snail	Terrestrial	25.00	16.27 [+/-93420.94]	8.51 [+/-2202.73]	619.54
	Oxystele antoni	Winkle	Aquatic	8.33	0.38 [+/-2345.49]	5.49[+/-1542.61]	48.86
	Oxystele tigrina	Winkle	Aquatic	8.33	0.03 [+/-180.42]	3.04 [+/-853.40]	25.54
	Scutellastra longicosta	Limpet	Aquatic	8.33	0.03 [+/-180.42]	0.03 [+/-9.02]	0.51
Echinodermata	*Parvulastra exigua*	Sea Star	Aquatic	8.33	0.23 [+/-1443.38]	2.06 [+/-579.15]	19.10

Cross-shore Zonation Patterns

Two types of patterns can be extracted from these data. Firstly, the zonation patterns of
individual species can be examined and secondly the overall patterns in species richness,
biomass and abundance across the shore can be examined.

Individual Species

The distribution patterns of each species across the shore from low to high tide levels are shown
in Table 3.3. As would be expected, species composition changed radically in response to tidal
levels. A small proportion of species showed a wide intertidal distribution, but over half of the
species were polarized to either the upper drift line areas of the shore (level 8 -7) or to the low
shore (levels 2-1). Of 12 species confined to the upper shore, only one Nemertea was aquatic.
The remaining 11 were air-breathing Arthropoda, including all Arachnida and Chilopoda, five of
six Insecta, three of five Isopoda and Amphipoda, plus one pulmonate Gastropoda. All 18
species confined to the low shore were aquatic forms, with most being forms typical of rocky
shore biota and uncommon in the samples. Only a single Syllidae and Nemertea, remained
confined in the vertical mid-shore zone. The remaining seven species showed wide distribution
patterns extending across most of the intertidal zone. Indeed, two species, *Perinereis namibia*
and *Ligia dilatata*, were reported across all eight tidal levels with *Procerodes sp.* following
closely behind occurring at seven of the eight tidal levels (Table 3.3).

Species Richness

Patterns of species richness across the shore levels are plotted in Figure 3.6. Interestingly, these did not show the sharp increases in species richness at lower shore level that is typical of other shore types. Instead, there is a 'u' shaped profile with highs of 18 species (level 8) occurring in the upper shore near the high drift line, resulting largely from the presence of many terrestrial air breathing taxa, and of 21 species (level 1) occurring in the lower shore near the water mark (Figure 3.6). The lowest species richness occurred in the mid shore zones (levels 5 and 4) which showed similarly low species richness of 5 - 7 species.

Table 3.3 Biomass (BM) and density (D) per taxa tabulated across the eight tidal levels pooled from all sites (n = 12) with 1 being the lowest and 8 being the highest level. Biomass (g/m²) and density (#/m²). Total mean biomass (g/m²) with standard deviations, total mean density (#/m²) with standard deviations and total species richness is reported for each tidal level.

Mean Biomass (g/m²)/Mean Density (#/m²) per Tidal Level		Tidal Level							
		Low							High
Species Accumulation		*1*	*2*	*3*	*4*	*5*	*6*	*7*	*8*
Group	*Scientific Name*	*BM/D*	*BMD*	*BM/D*	*BM/D*	*BM/D*	*BM/D*	*BM/D*	*BM/D*
Anthozoa	*Bunodosoma capense*	2.48/2.08							
Platyhelminthes	*Procerodes sp.*	0.02/14.58	0.02/20.83	0.02/10.42	0.02/2.08	0.04/8.33	0.04/50.00		0.02/4.17
Nemertea	*Cerebratulus fuscus*								0.13/2.08
	Nemertean unidentified		0.15/4.17		0.02/2.08				
Polychaeta	*Arabella iricolor*	3.38/4.17							
	Perinereis namibia	2.08/72.92	0.25/18.75	0.77/210.42	0.04/27.08	0.98/202.08	12.52/1868.75	0.13/35.42	0.15/41.67
	Pseudonereis variegata	0.96/2.08							
	Syllidae unidentified				0.02/4.17				
Arachnida	*Amaurobioides africana*							0.04/2.08	0.10/18.75
Chilopoda	*Geophilomorpha sp.*							0.02/4.17	0.02/2.08
Insecta	*Anurida maritima*		0.02/4.17	0.02/4.17		0.04/16.67	0.02/20.83	0.02/31.25	
	Cercyon maritimus								0.02/2.08
	Fucellia capensis								0.06/8.33
	Stratiomyidae (larvae)							0.02/2.08	0.08/12.50
	Staphylinidae unidentified								0.02/4.17
	Tabanidae (larvae)								0.73/2.08
Isopoda	*Deto echinata*						4.10/41.67	8.92/60.42	5.79/81.25
	Exosphaeroma truncatitelson	0.02/2.08	0.02/8.33						
	Ligia dilatata	0.02/4.17	17.81/1597.92	8.31/566.67	5.19/256.25	10.63/679.17	34.90/450.00	14.54/185.42	41.33/864.58

Group	Species								
	Marioniscus spatulifrons						1.04/14.58	7.04/83.33	19.27/204.17
	Philoscia sp.								0.02/12.50
Amphipoda	*Eorchestia dassenensis*			0.02/2.08				0.08/4.17	1.46/235.42
	Paramoera bidentata	0.02/2.08	0.10/20.83						
	Paramoera capensis	0.96/254.17	0.06/6.25	0.10/20.83	0.04/4.17		0.02/2.08		0.02/8.33
	Ptilohyale plumulosa	0.02/2.08		0.02/4.17			0.06/6.25	0.25/54.17	
	Talorchestia capensis			0.02/2.08	0.02/2.08	0.02/6.25	3.04/222.92	1.23/27.08	1.06/104.17
Bivalvia	*Aulacomya atra*	8.98/2.08							
Polyplacophora	*Acanthochitona garnoti*	1.35/2.08	1.54/2.08						
Gastropoda	*Burnupena cincta*	12.69/2.08							
	Burnupena lagenaria	9.54/2.08							
	Cymbula miniata	16.58/2.08							
	Cymbula oculus	1.21/2.08							
	Helcion pectunculus		0.23/2.08						
	Helcion pruinosus	11.29/35.42	6.88/14.58						
	Myosotella myosotis							27.60/1170.83	0.58/43.75
	Oxystele antoni	8.44/12.50	9.38/14.58						
	Oxystele tigrina	9.85/2.08							
	Scutellastra longicosta	0.10/2.08							
Echinodermata	*Parvulastra exigua*	6.69/16.67							
TOTAL Mean Biomass (g/m^2) +/- SD / Mean Density (#/m^2) +/- SD per Tidal Level		96.69 [+/-215.79]/ 441.67 [+/-1025.54]	36.46 [+/-83.94]/ 1714.58 [+/-5502.26]	9.29 [+/-27.61]/ 820.83 [+/-1897.78]	5.35 [+/-17.45]/ 297.92 [+/-830.63]	11.71 [+/-33.37]/ 912.50 [+/-2288.89]	55.75 [+/-134.88]/ 2677.08 [+/-4346.57]	59.90 [+/-127.22]/ 1660.42 [+/-3729.10]	70.88 [+/-152.83]/ 1652.08 [+/-3218.07]
Species Richness		21	12	8	7	5	9	12	18

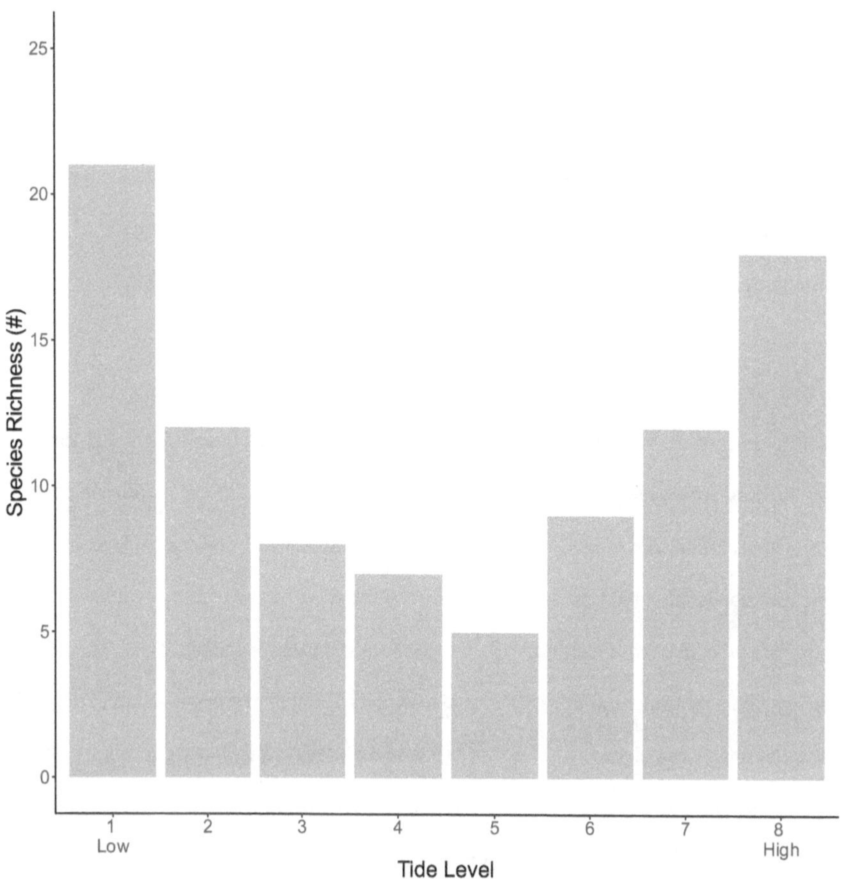

Figure 3.6. Species richness across each of the 8 tidal levels from low to high shore with level 1 being the lowest and 8 being the highest.

Biomass

At a species by species level, only six species ever attained mean biomass levels >10 g/m². The Isopoda, *L. dilatata,* had by far the highest biomass exceeding 10 g/m² at five of the eight tidal levels. The other five species only reached that biomass at a single tidal level, namely level 7 by the pulmonate Gastropoda, *M. myosotis,* level 6 by the Polychaeta, *P. namibia,* and level 1 by the Gastropoda, *Burnupena cinta, Cymbula miniata* and *Helcion pruinosus* (Table 3.3).

In terms of trends in mean total biomass across the shore, the high shore tidal levels, closest to the drift line, and the low shore tidal levels, closest to the water mark had similarly high mean total biomass, with level 7 at 96.69 g/m², and level 1, at 70.88 g/m² (Table 3.3) with level 1 having upper boundaries of the 75th quartile >2g/m² (Figure 3.7). The upper shore levels had the most variation in mean biomass due to outliers with level 8 having the greatest variation with one of its outliers, *L. dilatata,* exceeding >40 g/m², the second greatest variation, also due to, *L. dilatata* with a mean biomass of 34.90 g/m², was at level 6, and third greatest at level 7 due to *M. myosotis* with a mean biomass of 27.60 g/m² (Figure 3.7). Mean total biomass declined towards the mid shore levels 3 - 5 with tidal level 4 having the lowest mean biomass of 5.35 g/m² (Table 3.3) with an upper boundary <1g/m² (Figure 3.7).

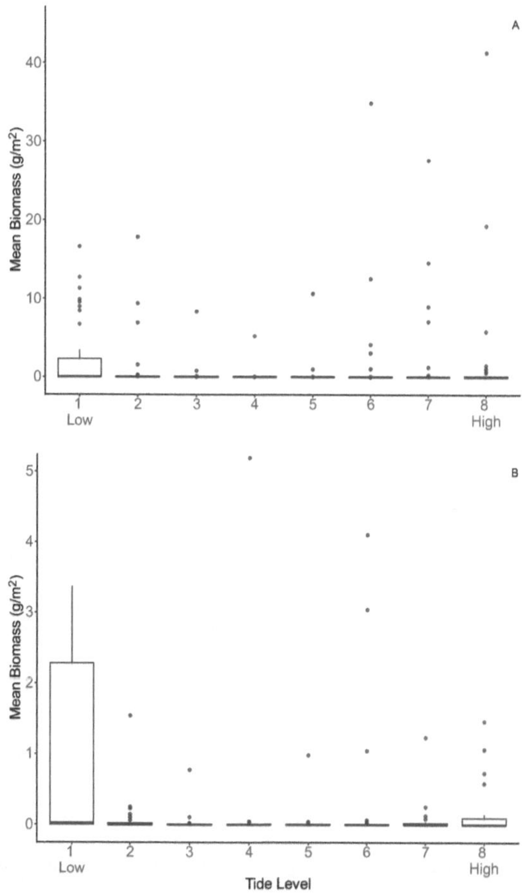

Figure 3.7 Box-plot of mean total biomass across each of the eight tidal levels from low to high shore, with 1 being the lowest and 8 being the highest level, expressed as g/m². Lower and upper box boundaries are 25th and 75th quartiles, the line inside the box is the median, with upper and lower error bars displaying 10th and 90th percentiles and filled dots as outliers falling outside of these percentiles, respectively. A is full view and B is zoomed view.

Density

The mean densities of each species across the tidal levels are also displayed in Table 3.3. The only three species which ever attained mean densities exceeding 1000 per square meter (#/m²) were *P. namibia at* 1868.75 #/m² at level 6, *L. dilatata* at 1597.92 #/m² at level 2, and *M. myosotis* 1170.83 #/m² at level 7 (Table 3.3). By far the most abundant species overall was *L. dilatata*, which attained densities >100 #/m² at seven of the eight tidal levels. Only two other species attained this density at more than one level, these being *P. namibia* (three levels) and *Talorchestia capensis* (two levels).

The pattern for total density was erratic, with highest density occurring at tidal level 6 at 2677.08 #/m² (Table 3.3). This was followed by the low shore tidal level 2 at 1714.58 #/m² and the similarly dense high shore levels 7 at 1660.42 #/m² and 8 at 1652.08 #/m². The lowest density occurred in the mid shore at tidal level 4 at 297.92 #/m² (Table 3.3). However, total densities at specific levels tended to be dominated by very high counts for individual species at specific levels, for example the peak at level 6 is driven largely by extremely high densities of *P. namibia* at 1868.75 #/m² and at level 2 by *L. dilatata* at 1597.92 #/m² and level 7 by *M. myosotis* at 1170.83 #/m² (Table 3.6). These trends are reflected in Figure 3.8 with the high shore (7) and low shore tidal levels (1 and 2) having the highest upper boundary 75th quartiles >2#/m² with tidal level 8 having the highest upper boundary 75th quartile >5#/m². The greatest variation of mean density (#/m²) occurred at tidal level 6 having the greatest variation with highest upper boundary 75th quartile <1 #/m² and the outlier, *P. namibia,* at <2000 #/m² (Figure 3.8).

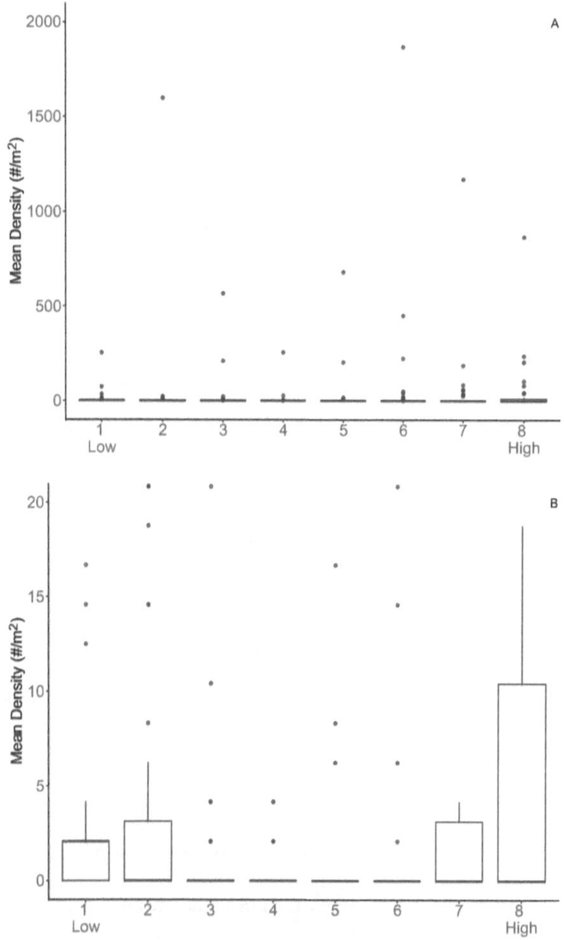

Figure 3.8 Box-plot of mean total abundance across each of the eight tidal levels from low to high shore, with 1 being the lowest and 8 being the highest level, expressed per meter squared (#/m²). Lower and upper box boundaries are 25th and 75th quartiles, the line inside the box is the median, with upper and lower error bars displaying 10th and 90th percentiles and filled dots as outliers falling outside of these percentiles, respectively. A is full view and B is zoomed view.

Discussion

Based on Chapter 2 it was shown that intermediate shores with particle sizes ranging from 1 – 256mm represent a separate habitat type, distinct from either sandy shores or those composed of boulders and solid rock. Focusing in on shores only within this grouping it is further demonstrated here that pebble and cobble shores form a single grouping in nMDS plots (Figure 3.4) and in the cluster dendrogram (Figure 3.5). It is thus concluded that pebble and cobble shores comprise a single habitat type with a similar and distinct biota. The aim of this chapter is to better quantify the fauna of pebble and cobble shores, by investigating the taxonomic composition of the fauna and examining their intertidal distribution patterns.

Abundance and Location of Pebble and Cobble Shores

According to Jackson and Lipschitz (1984) *'Coastal Sensitivity Atlas of Southern Africa 1984,'* very few pebble shores occur in the Western Cape with none listed in the Cape Agulhas to Jacobsbaai study area. Pebble and cobble sites thus had to be found before sampling could occur and this study demonstrates for the first time that pebble and cobble shores do indeed exist in this region (Cape Agulhas to Jacobsbaai). However, it is doubtful that all of the pebble and cobble shores within the study area have been found and a proper mapping effort is still required (and is proposed outside the scope of this dissertation). Furthermore, pebble and cobble shores were sampled throughout the study area; up the West Coast, along the Cape Peninsula and in False Bay and to its east, and despite geography the same biotic composition and trends in cross-shore zonation patterns were found.

Overall Composition of the Biota

A total of 39 species were found on the 12 pebble and cobble shores surveyed, much fewer than

Tucker (2017) found on either rock (124) or boulder (175) shores in the same region and with

similar sampling effort. This can be attributed to the harsh nature of this habitat, which is too

mobile for typical sedentary boulder or rocky shore forms to attach to. Of these 39 pebble and

cobble shore species, 14 were air breathing 'terrestrial' forms. This is unusual, as most intertidal

habitats are dominated by aquatic marine species, with air breathing forms making up only a

small proportion of the fauna, mostly confined to the upper tidal reaches (Branch & Branch,

2018). Of these 39 species found two-thirds (26) were encountered only once, suggesting that

they are not typical of this habitat, indeed the majority of these were typical rocky shore forms

(limpets, mussels, anemones etc) that occupied single larger rocks located, amongst much

smaller stones, in the lower reaches of one of a few of the shores. Therefore, only the remaining

14 species are considered truly characteristic of pebble and cobble shores (Appendix C, D, and

E).

Taxonomically the biota also lacked taxonomic diversity, the majority of species being either

Arthropoda (18) or Gastropoda (10). This is typical of other international studies on pebble and

cobble shores such as an Antarctic study (Jażdżewki et al., 2001) which also found mostly

amphipods, gastropods and nemerteans on stony shores. There was a notable absence of both

algae, and of numerous other major taxa that form significant components of the biota of both

rock and boulder shores, for example Porifera, Bryozoa, and Ascidiacea. Another interesting

aspect of the faunal composition was that species with the highest IRI values (Table 3.3) were

mostly terrestrial air breathers found in a variety of shore types. These included drift line mobile

detritus feeders, like the Isopoda, *L. dilatata* and *D. echinata*, that are typically associated with rocky or boulder shores, but also the drift line Amphipoda, *T. capensis,* which is normally associated with sandy beaches, the Polychaeta, *P. namibia,* typical of mixed shores and the introduced pulmonated Gastropoda, *M. myosotis*, an introduced species that has only previously been recorded from estuaries and freshwater seepages (Herbert, 2012). Interestingly, although rocky shores in this region are heavily invaded by introduced species, none of these were recorded in the study.

The low species richness on cobble and boulder shores can be attributed to the extremely harsh nature of the environment, the constant mobility of tumbling stones making up the substratum creating an unsuitable habitat for attached species including algae (Liberman, 1979), which in turn means that grazers are largely absent. As a result, to escape the heavy wave action and harsh conditions during high tide, air breathing terrestrial inhabitants certainly migrate up above the intertidal during high tide to escape the extremely harsh conditions caused by wave action, which results in tumbling of stones making up the substratum. Similar migratorybehaviour in these species has been noted on sandy or solid rock shores and is associated with scavenging for drift kelp in these habitats (Koop & Field, 1980, 1981; Griffiths, & Stenton-Dozey, 1981; Koop & Griffiths, 1982, Koop et al., 1982). Other mobile aquatic species probably survive extremem wave action at high tide by migrating deeper into the pebble/cobble bed, or in the case of the few larger attached forms recorded (the anemone, *B. capense,* the bivalve, *A. atra,* the chiton, *A. garnoti,* and the limpets, *C. miniata, C. oculus, H. pectunculus, H. pruinosus and S. longicornis*) remain confined to a few larger stones low on the shore.

Cross-shore Zonation Patterns

The distribution pattern for the biota of South African pebble and cobble shores included a species diverse upper shore mostly comprised of terrestrial inhabitants, with species numbers declining through the mid shore and increasing once more at the lower shores, mostly comprised of aquatic inhabitants typical of rocky shores. Biomass and density trends followed similar 'u' shaped trends with increases in both biomass and density near the high (6 - 8) and low (1 and 2) shore levels. Usually across sandy and rocky shores, species richness increases lower on the shore and closer to the water-mark due to more benign physical conditions (McLachlan, 1996; Tucker et al., 2017). The situation reported here is thus unusual when looking at shore types however, it is also replicated on reflective beaches (McLachlan, 1996; Short, 1996).

Air-breathing species high on the shore are able to escape the tumbling of the pebbles and cobbles caused by wave action by remaining out of the water and are able to shelter in the interstices amongst the grains (Fanini et al., 2019), where they can obtain a rich source of nutrition in the form of washed up plant material (Barber 2009). Shores in this region are often surrounded by dense forests of kelp, which become uprooted and deposited either on the high shore near the drift line (Koop & Field, 1980, 1981; Griffiths, & Stenton-Dozey, 1981; Koop & Griffiths, 1982, Koop et al., 1982), or collect as detritus within grain interstices at lower levels, thus providing nutrition to species lower on the shore.

It is interesting to note that many of the species recorded, including airbreathing forms, ranged well into the lower reaches of the shore during low tide. Indeed *L. dilatata* occurred in all eight sampled zones, right down to the low water mark, contributing the highest biomass exceeding 10 g/m^2 at five of the eight tidal levels, as well as being the only species to exceed >100 $\#/m^2$ at seven of eight tidal levels. According to Koop and Field (1980 and 1981), *L. dilatata* is a scavenger whose diet is predominately drift kelp, as such it is very likely they migrate up and down the shore for feeding purposes (Weatherington, 2014).

Conclusion

Overall Aim – Chapter 2

Chapter 2 attempted to compare the full spectrum of particle grain size shores including rocky

shores, and those categorized by the particle grain size classes identified in the Wentworth

(1922) scale, fine sand through boulder shores. Biological and physical data comprised of

sampled surveys from 16 sites and extracted data from 42 previously sampled sites occurring

between Cape Agulhas and Jacobsbaai were converted to uniform units and compared.

Comparisons were done to determine the overall similarity of shore types to one another as well

as the biotic composition of and species ranges across shore types.

Main Outcomes – Chapter 2

- nMDS and dendrogram clusters revealed three groupings at 10 % similarity amongst
  shore types, sandy (<1mm), intermediate particle grain sizes (1 – 256mm) and boulder
  and rocky (>256mm).
- Species presence polarized to shores of smaller (<1mm) and larger (>256mm) particle
  grain sizes as the rough conditions of tumbling grain that occur on shores of intermediate
  particle sizes (1 – 256mm) allow for few species to inhabit these shore types.
- Dramatic increases in species richness (#), biomass (g/m²) and density (#/m²) were seen
  in larger particle grain size shores (4 – 256mm) with the lowest values found on
  intermediate particle size shores (1 – 4mm).
- SIMPER results showed mobile species contributing the highest percentage to particle
  grain size shores 0.125 – 256mm with mobile burrowers preferring fine sand and medium

sand shore types (<0.5mm) while species contributing the highest percentage to larger

particle grain sizes (>256mm) tended to be attached sessile animals.

Overall Aim – Chapter 3

Chapter 3 attempted to use 12 of the sampled sites from Chapter 2, five cobble and seven pebble

shores, to quantify the unique fauna of pebble and cobble shores and explore their distribution

patterns.

Main Outcomes – Chapter 3

- Pebble (4 –< 64mm) and cobble (64 –< 256mm) shore types had similar species presence

  and composition, and all sites were compared using fourth root transformed density

  (#/m²) nMDS and cluster analysis revealing no groupings amongst sites, suggesting they

  form a single habitat type.

- Taxa occupying pebble and cobble shores (4 – 256mm) were mostly comprised of

  terrestrial air-breathing scavengers who concentrate on the high shore and are thought to

  migrate to the low shore coinciding with low tide in search of deposited detritus,

  especially drift kelp.

- *L. dilatata* had the highest IRI value, was found on all eight tidal levels and by far

  contributed the highest biomass exceeding 10 g/m² at five of eight tidal level, as well as

  being the only species to exceed >100 #/m² at seven of eight tidal levels, further

  supporting the migrating scavenger argument.

Study Limitations

Limitations of the study mostly involved constraints in the amount of sampling efforts that could be undertaken. Previously sampled sites should ideally have been re-sampled using our own standardized methods, for a variety of reasons. Firstly, some studies were conducted in the previous century and physical characteristics and fauna may have changed in the several decades since these earlier studies were conducted. New environmental factors may also be influencing species composition on those sites for example the creation of the West Coast National Park as well as the construction of the Saldanha Bay harbour. Beyond this, differing sampling effort between studies may also affect results, for example the more diverse rocky shores present in Day's (1959) prolonged study as opposed to the less diverse rocky shores present in Tucker et al.'s (2017) single transect study. Apart from various methods used, taxonomic abilities of previous authors may also vary, as many were students at the time when their studies were undertaken with consulted experts finding misidentifications among this work (Puttick, 1977; Bally, 1981, 1987; Nel, 2000; Soares, 2003; Tucker, 2014; Tucker et al., 2017; Harris per comms) while other authors were well practiced taxonomists (Day, 1959; Brown, 1971). Lastly, different authors used varying methods to measure grain size and faunal biomass. Grain size was either given as median (Day, 1959; Brown, 1971; Puttick, 1977) or mean (Bally, 1987; Nel, 2000; Soares, 2003; Tucker, 2014; Harris pers comm) particle size. However, all authors used the Wentworth (1922) scale, which allowed for accurate shore type classification. Biomass was given as dry weight, acidified dry weight, ash free dry weight, wet weight as well as further variants regarding whether or not shells were retained. As such, all of biomass had to be made uniform using published conversions (Field et al., 1980; Riccardi & Bourget,1998). Due to restrictions of the pandemic and the successive lockdown, such field outings to conduct our own surveys using uniform methods on these historical sites could not be completed. All cobble

shores found within the study area were sampled, however, more cobble shores should ideally

have been sampled allowing the accumulation curve to reach an asymptote. Expanding the study

area may have resolved this. Ideally shores across all four bioregions, especially in the Natal and

Delgoa bioregions, should have been sampled so that biogeographic differences in faunal

composition could be included in the study. Another, more appropriate, option, considering this

is a Masters level project, would have been to undertake more concentrated surveys of fewer

shores. Lastly, the abundance and biomass representations in this study of mobile fauna

occupying larger particle grain size shores, such as pebbles (4 –< 64mm) and cobbles (64 –<

256mm), could be argued inaccurate as many animals find easy escapes within grain interstices

when sampling. Thus, not all animals originally present within a quadrat when originally lain are

collected once the grain within the quadrat is disturbed.

Further Research

Further research on these shore types would include using ground-truthing the Jackson and

Lipschitz (1984) *'Coastal Sensitivity Atlas of Southern Africa 1984'* to map the shore types along

the study area. Efforts to do so were originally planned for this dissertation, however due to

lockdown restrictions in response to the global pandemic those efforts were delayed. Current

conversations with SAEON are underway to resurrect this work as a separate project, however,

due to study visa limitations, this work could not be included in the dissertation. Beyond

mapping, further work to sample the remaining three bioregions for coarse sand through boulder

shores (0.5 – 256mm) is recommended, this would allow insight into whether or not similar

trends are found in other bioregions under different environmental conditions. This would also allow more work to be done on the already identified and understudied habitat type of boulder shores, which currently has been the subject of just one paper published in South Africa (Tucker et al., 2017). Lastly, exploring the migratory behaviour patterns of pebble and cobble shore inhabitants would prove interesting, as unlike other shore types, biota here are mainly air-breathing terrestrial scavengers concentrated on the high shore. Such a unique intertidal ecosystems should be investigated for new ecological trends and conservation efforts for these shore types.

References

Asano, T., Sato, S., Lui, H., & Takagawa, T. (2010). Experimental study on effect and advantage of beach nourishment using coarse sand. *Journal of Japan Society of Civil Engineers, 66*(1), 631-635.

Apolinário, M. (1999). The role of pre-recruitment processes in the maintenance of a barnacle (*Chthamalus challengeri* Hoek) patch on an intertidal pebble shore in Japan. *Revista Brasil Biologia, 59*(2), 225-237. https://doi.org/10.1590/S0034-71081999000200007

Auguie, B. (2017). gridExtra: Miscellaneous Functions for "Grid" Graphics. R package version 2.3. https://CRAN.R-project.org/package=gridExtra

Awad, A. A., Griffiths, C. L., & Turpie, J. K. (2002). Distribution of South African marine benthic invertebrates applied to the selection of priority conservation areas. *Diversity and Distributions, 2002*(8), 129-145. Retrieved 2020, from http://www.blackwell-science.com/ddi

Bally, R. (1981). *The ecology of three sandy beaches on the west coast of South Africa* [Doctoral Dissertation, University of Cape Town, South Africa]. Retrieved from http://hdl.handle.net/11427/15452

Bally, R. (1987). The ecology of sandy beaches of the Benguela ecosystems. *South African Journal of Marine Science, 5*(1), 759-770. https://doi.org/10.2989/025776187784522685

Barber, A. D. (2009). Littoral myriapods: A review. *Soil Organisms, 81*(3). 735-760.

Barnes, D. K. A., & Lehane, C. (2001). Competition, mortality, and diversity in South Atlantic coastal boulder communities. *Polar Biology, 24*(3), 200-208. https://doi.org/10.1007/s003000000196

Blamey, L. K., & Branch, G. M. (2009). Habitat diversity relative to wave action on rocky shores: Implications for the selection of marine protected areas. *Aquatic Conservation: Marine and Freshwater Ecosystems, 19*(6), 645-657. https://doi.org/10.1002/aqc.1014

Blott, S. J., & Pye, K. (2001). Gradistat: A grain size distribution and statistics package for the analysis of unconsolidated sediments. *Earth Surface Processes and Landforms, 26*(11), 1237-1248. https://doi.org/10.1002/esp.261

Branch, G. M. (2001). Rocky shores. In J. H. Steele, S. A. Thorpe & K. K. Turekian (Eds.), *Encyclopedia of ocean science* (pp. 2427-2434). Academic Press.

Branch, G. M., & Branch, M. L. (2018). *Living shores: Interacting with southern Africa's marine ecosystems* (E. du Plessis & C. deVilliers, Eds.). Struik Nature, Cape Town. (Original work published 1981)

Branch, G. M., & Griffiths, C. L. (1988). The Benguela ecosystem: Part V. The coastal zone. *Oceanography and Marine Biology: An Annual Review, 26,* 395-486.

Branch, G. M., Griffiths, C. L., Branch, M. L., Beckley, L. E. (2010). *Two oceans: A guide to the marine life of southern Africa* (E. Bowles, Ed.). Struik Nature, Cape Town. (Original work published in 1994)

Brown, A. C. (1964). Food relations on the intertidal sandy beaches of the Cape Peninsula. *South African Journal of Science, 60,* 35-41.

Brown A. C. (1971). The ecology of sandy beaches of the Cape Peninsula, South Africa. Part I: Introduction. *Transactions of the Royal Society of South Africa, 39,* 247-279. https://doi.org/10.1080/00359197109519119

Brown, A. C. (1996). Behavioural plasticity as a key factor in the survival and evolution of the macrofauna on exposed sandy beaches. *Revista Chilena Historia Natural, 69,* 469-474. Retrieved 2020, from http://rchn.biologiachile.cl/pdfs/1996/4/Brown_1996.pdf

Brown, A. C., & Jarman, N. J. (1978). Coastal marine habitats. In Werger, M. J. A., (Ed.) *Biogeography and ecology of southern Africa,* (pp. 1239-1277). Dr. W. Junk, The Hague.

Brown, A. C. & McLachlan, A. (2002). Sandy shore ecosystems and the threats facing them: Some predictions for the year 2025. *Environmental Conservation, 29*(1), 62–77. https://doi.org/10.1017/S037689290200005X

Bujan, N., Cox, R., Lin, L., Ducrocq, C., & Hwung, H. (2018). Semiautomatic Digital Clast Sizing of a Cobble Beach, Nantian, Taiwan. *Journal of Coastal Research, 34*(6), 1367-1381. https://www.jstor.org/stable/26538615

Bujan, N., Cox, R., & Masselink, G. (2019). From fine sands to boulders: Examining the relationship between beach-face slope and sediment size. *Marine Geology, 417,* 1-17. https://doi.org/10.1016/j.margeo.106012

Bustamante, R. H., & Branch, G. M. (1996). Large scale patterns and trophic structure of South African rocky shores: The roles of geographic variation and wave exposure. *Journal of Biogeography, 23*(3), 339-351. Retrieved 2020, from http://www.jstor.org/stable/2845850

Chapman, M. G. (2002). Early colonization of shallow subtidal boulders in two habitats. *Journal of Experimental Marine Biology and Ecology, 275*(2), 95-116. https://doi.org/10.1016/S0022-0981(02)00134-X

Chapman, M. G. (2003). The use of sandstone blocks to test hypotheses about colonization of intertidal boulders. *Journal of the Marine Biological Association of the United Kingdom, 83*(2), 415-423. https://doi.org/10.1017/S0025315403007276h

Chapman, M. G. (2005). Molluscs and echinoderms under boulders: Test of generality of patterns of occurrence. *Journal of Experimental Marine Biology and Ecology, 325*(1), 65-83. https://doi.org/10.1016/j.jembe.2005.04.016

Chapman, M. G. (2008). Patterns of spatial and temporal variation of macrofauna under boulders in a sheltered boulder field. *Austral Ecology, 27*(2), 211-228. https://doi.org/10.1046/j.1442-9993.2002.01172.x

Chapman, M. G. (2012). Restoring intertidal boulder-fields as habitat for 'specialist' and 'generalist' animals. *Restorative Ecology, 20*(2), 277-285. https://doi.org/10.1111/j.1526-100X.2011.00789.x

Chapman, M. G. (2013). Constructing replacement habitat for specialist and generalist molluscs - the effect of patch size. *Marine Ecology Progress Series, 473,* 201-214. https://doi.org/10.3354/meps10074

Chapman, M. G. (2017). Intertidal boulder-fields: A much neglected, but ecologically important, intertidal habitat. *Oceanography and Marine Biology: An Annual Review, 2017*(55), 35-54. https://doi.org/10.1111/j.1526-100X.2011.00789.x

Chapman, M. G. & Underwood, A. J. (1996). Experiments on effects of sampling on biota under intertidal and shallow subtitle boulders. *Journal of Experimental Marine Biology and Ecology, 207*(1-2), 103-126.

Clarke, K. R. (1993). Non-parametric multivariate analyses of changes in community structure. *Australian Journal of Ecology, 18*(1), 117-143. https://doi.org/10.1111/j.1442-9993.1993.tb00438.x

Clarke K. R., & Gorley R N. (2006). PRIMER (Plymouth Routines in Multivariate Ecological Research) v6: *User manual/tutorial 1ˢᵗ edition, pp190.* Plymouth, UK: PRIMER-E

Connell, J. H. (1978). Diversity in tropical rain forests and coral reefs. *Science, 199*(4335), 1302-1310. https://doi.org/10.1126/science.199.4335.1302

Cunha, P. L., & Cunha, M. A. (2016, September 5-9). *Index of Relative Importance; a new perspective* [Oral Presentation]. Frontiers in Marine Science Conference Abstract: XIX Iberian Symposium on Marine Biology Studies, Porto, Portugal. https://doi.org/10.3389/conf.FMARS.2016.05.00038

Davy, A. J., Willis, A. J., & Beerling, D. J. (2001). The plant environment: aspects of the ecophysiology of shingle species. In J. R. Packham, J. M. Randall, R. S. K. Barnes & A Neal (Eds.), *Ecology and geomorphology of coastal shingle* (pp. 191-201). Otley, West Yorkshire, Westbury. https://doi.org/10.1002/esp.379

Day, J. H. (1959). The biology of Langebaan Lagoon: A study of the effect of shelter from wave action. *Transaction of the Royal Society of South Africa, 35*(5), 475-547. https://doi.org/10.1080/00359195909519025

Defeo, O., & McLachlan, A. (2005). Patterns, processes and regulatory mechanisms in sandy beach macrofauna: A multi-scale analysis. *Marine Ecology Progress Series, 295,* 1-20. https://doi.org/10.3354/meps295001

Defeo, O., & McLachlan, A. (2013). Global patterns in sandy beach macrofauna: Species richness, abundance, biomass and body size. *Geomorphology, 199,* 106–114. https://doi.org/10.1016/j.geomorph.2013.04.013

Defeo, O., McLachlan, A., Schoeman, D. S., Schlacher, T. A., Dugan, J., Jones, A., Lastra, M., & Scapini, F. (2009). Threats to sandy beach ecosystems: A review. *Estuarine, Coastal and Shelf Science, 81*(1), 1–12. https://doi.org/10.1016/j.ecss.2008.09.022

de la Huz, R., Lastra, M., & López, J. (2002). The influence of sediment grain size on burrowing, growth and metabolism of *Donax trunculus* L. (Bivalvia: Donacidae). *Journal of Sea Research, 47*(2), 85-95. https://doi.org/10.1016/S1385-1101(02)00108-9

Dexter, D. M. (1992). Sandy beach community structure: The role of exposure and latitude. *Journal of Biogeography, 19*(1), 59-66. https://doi.org/10.2307/2845620

Dinno, A. (2017). dunn.test: Dunn's Test of Multiple Comparisons Using Rank Sums. R package version 1.3.5. https://CRAN.R-project.org/package=dunn.test

Dye A. H. (1988). Rocky shore surveillance on the Transkei Coast, Southern Africa: Temporal and spatial variability in the balanoid zone at Dwesa. *South African Journal of Marine Science, 7*(1), 87-99. https://doi.org/10.2989/025776188784379143

Dye, A. H., McLachlan, A., & Wooldridge, T. (1981). The ecology of sandy beaches in Natal, South Africa. *South African Journal of Zoology, 16,* 200-209.

Fanini, L., Coleman, C. O., & Lowry, J. K. (2019). Insights into the ecology of *Cryptorchestia garbinii* on the shores of the urban lake Tegel (Berlin, Germany). *Life and Environment, 69*(2-3), 187-191.

Field, J. G. & Griffiths, C. L. (1991). Littoral and sublittoral ecosystems of southern Africa. In A. C. Mathieson & P. H. Nienhuis (Eds.), *Intertidal and littoral ecosystems (Ecosystems of the World volume 24).* (pp. 323-346). Elsevier Science Publisher, Amsterdam.

Field, J. G., Griffiths, C. L., Griffiths, R. J., Jarman, N., Zoutendyk, P., Velimirov, B., & Bowes, A. (1980). Variation in structure and biomass of kelp communities along the south-west Cape coast. *Transactions of the Royal Society of South Africa, 44,* 145-203.

Fiora, S. M., & Carcedo, M. C. (2015). Influence of grain size on burrowing and alongshore distribution of the yellow clam (*Amarilladesma mactroides*). *Journal of Shellfish Research, 34*(3), 785-789. https://doi.org/10.2983/035.034.0307

Forgie, S. A., St. John, M. G., & Wiser, S. K. (2012). Invertebrate communities and drivers of their composition on gravel beaches in New Zealand. *New Zealand Journal of Ecology, 37*(1), 95-104.

Fuller, R. M. (1987). Vegetation establishment on shingle beaches. *Journal of Ecology, 75*(4), 1077-1089. https://doi.org/10.2307/2260315

Furota, T., & Ito, T. (1999). Life cycle and environmentally induced semelparity in the shore isopod *Ligia cinerascens* (Ligidae) on a cobble shore along Tokyo Bay, Central Japan. *Journal of Crustacean Biology, 19*(4), 752-761. Retrieved 2020, from https://www.jstor.org/stable/a549299

Gauci, M. J., Deidun, A., & Schembri, P. J. (2005). Faunistic diversity of Maltese pocket sandy and shingle beaches: Are these of conservation value? *Oceanologia, 47*(2), 219-241.

Griffiths, C. L. (2005). Coastal marine biodiversity in East Africa. *Indian Journal of Marine Sciences, 34*(1), 35-41.

Griffiths, C. L., Robbins, A. E. (2019). Box 2. Boulder and cobble shores: Little-known coastal ecosystem types. In L. R. Harris, K. Sink, A. Skowno, & L. van Niekerk (Eds.), *South African national biodiversity assessment 2018: Technical report. Volume 5: Coast.* South

African National Biodiversity Institute, Pretoria. Retrieved 2020, from
https://www.sanbi.org/wp-content/uploads/2019/10/NBA-Report-2019.pdf

Griffiths, C. L., Robinson, T. B., Lange, L. & Mead, A. (2010). Marine biodiversity in South
Africa: An evaluation of currents states of knowledge. *Plos One, 5*(8), 1-13.
https://doi.org/10.1371/journal.pone.0012008

Griffiths, C. L., & Stenton-Dozey, J. (1981). The fauna and rate of degradation of stranded kelp.
Estuarine Coastal and Shelf Science 12(6), 645-653. https://doi.org/10.1016/S0302-
3524(81)80062-X

Harris, L. R., Campbell, E., Nel, R., & Schoeman, D. (2014). Rich diversity, strong endemism,
but poor protection: Addressing the neglect of sandy beach ecosystems in coastal
conservation planning. *Diversity and Distributions 20*(10), 1120-1135.
https://doi.org/10.1111/ddi.12226

Harris, L., Nel, R., Holness, S., Sink, K., & Schoeman, D. (in press) Setting conservation targets
for sandy beach ecosystems. *Estuarine Coastal and Shelf Science.*
https://doi.org/10.1016/j.ecss.2013.05.016

Harris, L., Nel, R., & Schoeman, D. (2011). Mapping beach morphodynamics remotely: A novel
application tested on South African sandy shores. *Estuarine Coastal and Shelf Science,
92*(1), 78-89. https://doi.org/10.1016/j.ecss.2010.12.013

Harris, L. R., Sink, K., Skowno, A., & van Niekerk, L. (2019). *South African national
biodiversity assessment 2018: Technical report. Volume 5: Coast.* South African National
Biodiversity Institute, Pretoria. Retrieved 2020, from https://www.sanbi.org/wp-
content/uploads/2019/10/NBA-Report-2019.pdf

Henninger, T. O., & Hodgson, A. N. (2000). Foraging activity of *Helcion pruinosus*
(Patellogastropoda) on a South African boulder shore. *Journal of Molluscan Studies,
67*(1), 59-68. https://doi.org/10.1093/mollus/67.1.59

Henninger, T. O., & Hodgson, A. N. (2001). The reproductive cycle of *Helcion pruinosus*
(Patellogastropoda) on two South African boulder shores. *Journal of Molluscan
Studies, 67*(3), 385-394. https://doi.org/10.1093/mollus/67.3.385

Herbert, D. G. (2012). *Myosotella myosotis* (Mollusca: Ellobiidae) – An overlooked, but well-
established introduced species in South Africa. *African Journal of Marine Science, 34*(3),
459-464. https://doi.org/10.2989/1814232X.2012.716374

Huggett, J., & Griffiths, C. L. (1986). Some relationships between elevation, physico-chemical variables and biota of intertidal rock pools. *Marine Ecology Progress Series 29*(2), 189-197. Retrieved 2020, from https://www.jstor.org/stable/24817587

Jackson, L. F. (1976). Aspects of the intertidal ecology of the east coast of South Africa. *Investigative Report Oceanographic Research Institute, Durban 46*, 1-72.

Jackson L. F., & Lipschitz, S. (1984). *Coastal sensitivity atlas of Southern Africa 1984.* Compiled for the Dept. of Transport of the Republic of South Africa, Cape Town.

Jażdżewski, K., De Broyer, C., Pudlarz, M., & Zielinski, D. (2001). Seasonal fluctuations of vagile benthos in the uppermost sublittoral of a maritime Antarctic fjord. *Polar Biology 24*(12), 910-917. https://doi.org/10.1007/s003000100299

Koop, K., & Field, J. G. (1980). The influence of food availability on population dynamics of a supra-littoral isopod, *Ligia dilatata* Brandt. *Journal of Experimental Marine Biology and Ecology, 48*(1), 61-72. https://doi.org/10.1016/0022-0981(80)90007-6

Koop, K., & Field, J. G. (1981). Energy transformation by the supra-littoral isopod *Ligia dilatata* Brandt. *Journal of Experimental Marine Biology and Ecology, 53*(2-3), 221-233. https://doi.org/10.1016/0022-0981(81)90021-6

Koop, K., & Griffiths, C. L. (1982). The relative significance of bacteria, meio- and macrofauna on an exposed sandy beach. *Marine Biology, 66*(3), 295-300.

Koop, K., Newell, R. C., & Lucas, M. I. (1982). Biodegradation and carbon flow based on kelp *(Ecklonia maxima)* debris in a sandy beach microcosm. *Marine Ecology Progress Series, 7*(3), 315-326. https://www.jstor.org/stable/24814559

Kurihara, T. (2001). Spatial and temporal fluctuations in densities of gastropods and bivalves on subtropical cobbled shores. *Bulletin of Marine Science -Miami, 68*(3), 409-426.

Kurihara, T. (2002). Spatial and temporal fluctuation in the density of the intertidal limpet, *Patelloida striata* Quoy & Gaimard, on subtropical cobbled shores. *Journal of Molluscan Studies, 68*(1), 79-86. https://doi.org/10.1093/mollus/68.1.79

Kurihara, T. (2007). Life-history traits of a gastropod, *Nerita squamulata* Le Guillou 1841, on a subtropical cobbled shore disturbed by sand. *Plankton and Benthos Research 2*(4), 213-218. https://doi.org/10.3800/pbr.2.213

Lasiak, T., & Field, J. G. (1995). Community-level attributes of exploited and non-exploited rocky infratidal macrofaunal assemblages in Tranksei. *Journal of Experimental Marine Biology and Ecology, 185*(1), 33-53. https://doi.org/10.1016/0022-0981(94)00130-6

LeHir, M., & Hily, C. (2005). Macrofaunal diversity and habitat structure in intertidal boulder fields. *Biodiversity and Conservation 14*(1), 233-250. https://doi.org/10.1007/s10531-005-5046-0

Liberman, M. (1979). Ecology of subtidal algae on seasonally devasted cobble substrates off Ghana. *Ecology, 60*(6), 1151-1161. https://doi.org/10.2307/1936963

Mathers, K. L., Rice, S. P., & Wood, P. J. (2019). Predator, prey, and substrate interactions: The role of faunal activity and substrate characteristics. *Ecosphere 10*(1), 1-18. https://doi.org/10.1002/ecs2.2545

Matsumoto, H., Young, A. P., & Guza, R. T. (2019). Observations of surface cobbles at two southern California beaches. *Marine Geology, 419*(2020), 1-10. https://doi.org/10.1016/j.margeo.2019.106049

McGuinness, K. A. (1984). Species-area relations of communities on intertidal boulders: Testing the null hypothesis. *Journal of Biogeography 11*(5), 439–456. https://doi.org/10.2307/2844807

McGuinness, K. A. (1987a). Disturbance and organisms on boulders. I. Patterns in the environment and the community. *Oecologia, 71*, 409–419. https://doi.org/10.1007/BF00378715

McGuinness, K. A. (1987b). Disturbance and organisms on boulders. II. Causes in patterns of diversity and abundance. *Oecologia, 71*, 420–430. https://doi.org/10.1007/BF00378316

McGuinness, K. A. (1988). Short-term effects of sessile organisms on colonization of intertidal boulders. *Journal of Experimental Marine Biology and Ecology, 106*(2), 159-175. https://doi.org/10.1016/0022-0981(88)90053-6

McGuinness, K. A., & Underwood, A. J. (1986). Habitat structure and the nature of communities on intertidal boulders. *Journal of Experimental Marine Biology and Ecology 104*(1-3), 97–123. https://doi.org/10.1016/0022-0981(86)90099-7

McLachlan, A. (1977). Composition, distribution, abundance and biomass of the macrofauna and meiofauna of four sandy beaches. *Zoologica Africana, 12*(2), 279-306. https://doi.org/10.1080/00445096.1977.11447576

McLachlan, A. (1996). Physical factors in benthic ecology: effects of changing sand particle size on beach fauna. *Marine Ecology Progress Series, 131*, 205-217. https://doi.org/10.3354/meps131205

McLachlan, A., & Bates, G. C. (1983). Sandy beach ecology – workshop report. In A. McLachlan, & T. Erasmus (Eds.), *Sandy beaches as ecosystems* (pp. 569-572). Springer, Netherlands.

McLachlan, A., & Brown, A. C. (2006). *The ecology of sandy shores.* Academic Press, Burlington, MA, USA.

McLachlan, A., & Defeo, O. (2013). Global patterns in sandy beach macrofauna: Species richness, abundance, biomass and body size. *Geomorphology, 199*(2013), 106-114. https://doi.org/10.1016/j.geomorph.2013.04.013

McLachlan, A., & Defeo, O. (2017). *The ecology of sandy shores* (3rd edition). Academic Press. https://doi.org/10.1016/B978-0-12-809467-9.00001-1

McLachlan, A., & Dorvlo, A. (2005) Global patterns in sandy beach macrobenthic communities: Biological factors. *Journal of Coastal Research, 23*(5), 674–687. Retrieved 2020, from https://www.jstor.org/stable/4496122

McLachlan, A., Erasmus, T., Dye, A. H., Woolridge, T., Van der Horst, G., Lasiak, T. A., & McGwynne, L. (1981a). Sand beach energetics: An ecosystem approach towards a high energy interface. *Estuarine Coastal and Shelf Science, 13*(1), 11-25. https://doi.org/10.1016/S0302-3524(81)80102-8

McLachlan, A., Jaramillo, E., Donn, T. E., & Wessels, F. (1993). Sandy beach macrofauna communities and their control by the physical environments: A geographic comparison. *Journal of Coastal Research 15*, 27 – 38. Retrieved 2020, from https://www.jstor.org/stable/25735721

McLachlan, A., Lombard, H. W., & Louwrens, S. (1981b). Trophic structure and biomass distribution on two east Cape rocky shores. *South African Journal of Zoology 16*(2), 85-89. https://doi.org/10.1080/02541858.1981.11447738

McLachlan, A., Wooldridge, T., & Dye, A. H. Y. (1981c). The ecology of sandy beaches in southern Africa. *South African Journal of Zoology, 16*(4), 219-231.

McQuaid, C. D., & Branch, G. M. (1984). The influence of sea temperature, substratum and wave exposure on rocky intertidal communities: An analysis of faunal and floral biomass. *Marine Ecology Progress Series, 19,* 145-151.

McQuaid, C. D., & Branch, G. M. (1985). Trophic structure of rocky intertidal communities: Response to wave action and implications for energy flow. *Marine Ecology Progress Series 22*(2), 153-161.Retrieved 2020, from https://www.jstor.org/stable/24816962

Mehrshad, T., Bastami, K. D., M. Y., Foshtomi, (2017, October 30-31). *The role of the sediment conditions in shaping meiofauna spatial distribution in the shallow water of the south Caspian Sea* [Paper]. First International Conference on Oceanography for West Asia, Tehran, Iran.

Menge, B.A. & Branch, G.M. (2001). Rocky intertidal communities. In M. D. Bertness, S. Gaines, & M. E. Hay (Eds.), *Marine community ecology* (pp. 221-257). Sinauer Associates, Sutherland.

Miroshnikova, Y. M., & Neradovsky, Y.N. (2019, April 17-18). *Pebble beaches of Murmansk coast unique formations of the Kola Peninsula* [Paper]. IOP Conference Series: Earth and Environmental Science, Volume 302, 4th International Scientific Conference "Arctic: History and Modernity", Saint Petersburg, Russian Federation. https://doi.org/1088/1755-1315/302/1/012046

Nakaoka, Y., & Wada, K. (2017). Habitat preference of *Cyclograpsus pumilio* (Decapoda, Brachyura, Varunidae) inhabiting the upper intertidal limit of pebble shores. *Japanese Journal of Benthology. 72,* 12-15.

Nel, P. (2000). *Physical and biological factors structuring sandy beach macrofauna communities* [Unpublished doctoral dissertation]. Department of Zoology, University of Cape Town, South Africa. Retrieved from http://hdl.handle.net/11427/6146

Nhon, D. H., Nguyen, T. M. L., Lai, T. B. T., Nguyen, N. N., Nguyen, H. H., & Bui, M. Q. (2019). Mineral compositions and grain sizes of sediments in intertidal zone in Northern Vietnam. *Vietnam Journal of Marine Science and Technology, 19*(3a), 63-75. https://doi.org/10.15625/1859-3097/19/3A/14292

Packman, J. R., Randall, R. E., Barnes, R. S. K., & Neal A (Eds.). (2001). *Ecology and geomorphology of coastal shingle*. Otley, West Yorkshire, Westbury.

Pezy, J. P., Baffreau, A., & Dauvin, J. C. (2016, August 1-5). *Revisited Syllidae of the English Channel coarse sand communities* [Poster]. 12th International Polychaeta Conference, National Museum Wales, Cardiff. https://doi.org/10.13140/RG.2.2.32003.71208

Pubill, E., Abelló, P., Ramón, M., & Baeta, M. (2011). Faunistic assemblages of a sublittoral coarse sand habitat of the northwestern Mediterranean. *Scientia Marina, 75*(1), 189-196. https://doi.org/10.3989/scimar.2011.75n1189

Puttick, G. M. (1977). Spatial and temporal variations in intertidal animal distribution at Langebaan Lagoon, South Africa. *Transactions of the Royal Society of South Africa, 42,* 403-433. https://doi.org/10.1080/00359197709519925

Quantum GIS 3.10 (A Coruna), QGIS Development Team 2019, QGIS Geographic Information System, Open Source Geospatial Foundation Project, http://qgis.osgeo.org

R Core Team (2013). R: A language and environment for statistical computing. R Foundation for Statistical Computing, Vienna, Austria. URL http://www.R-project.org/.

Raffaelli, D., & Hawkins, S. (1996). *Intertidal Ecology.* Chapman & Hall, London. https://doi.org/10.1017/S002531540003410X

Randall, R.E. (2008). The shingle vegetation of the coastline of New Zealand: Nelson Boulder Bank and Kaitorete Spit. *New Zealand Geographer, 93*(1), 11-19. https://doi.org/10.1111/j.0028-8292.1992.tb00306.x

Riccardi, A., & Bourget, E. (1998). Weight-to-weight conversion factors for marine benthic macro-invertebrates. *Marine Ecology Progress Series, 163,* 245-251. https://doi.org/10.3354/meps171245

Rius, M., & Teske, P. (2011). A revision of the *Pyura stolonifera* species complex (Tunicata, Ascidiacea), with a description of a new species from Australia. *Zootaxa, 2754*(1). 27-40. https://doi.org/10.11646/zootaxa.2754.1.2

Ruessink, G., Brinkkemper, J. A., & Kleinhans, M. G. (2015). Geometry of wave-formed orbital ripples in coarse sand. *Journal of Marine Science and Engineering, 3*(4), 1568-1594. https://doi.org/10.3390/jmse3041568

Russell, R. J. (2006.) Where most grains of very coarse sand and fine gravel are deposited. *Sedimentology, 11*(1-2), 31-38. https://doi.org/10.1111/j.1365-3091.1968tb00838.x

Ryu, S., Jang, K., Choi, E., Kim, S., Song, S. J., Cho, H., Ryu, J., Kim, Y., Sagong, J., Lee, J., Yeo, M., Bahn, S., Kim, H., Lee, G., Lee, D., Choo, Y., Pak, J., Park, J., Ryu, J., …

Hwang, U.W. (2012). Biodiversity of marine invertebrates on rocky shores of Dokdo, Korea. *Zoological Studies, 51*(5). 710-726. Retrieved 2020 from http://zoolstud.sinica.edu.tw/Journals/51.5/710.pdf

Schlacher, T. A., Schoeman, D.S., Dugan, J., Lastra, M., Jones, A., Scaponi, F., & McLachlan, A. (2008). Sandy beach ecosystems: Key features sampling issues, management challenges and climate change impacts. *Marine Ecology, 29*(suppl. 1), 70-90. https://doi.org/10.1111/j.1439-0485.2007.00204.x

Schlacher, T. A., Schoeman, D. S., Dugan, J., Lastra, M., Jones, A., Scaponi, F., McLachlan, A., & Dfeo, O. (2007). Sandy beaches at the brink? *Diversity and Distibutions, 13*(5), 556-560. https://doi.org/10.1111/j.1472-4642.2007.00363.x

Scott, R. J., Griffiths, C. L., & Robinson, T.B. (2012). Patterns of endemicity and range restriction among southern African coastal marine invertebrates. *African Journal of Marine Science, 34*(3), 341–347. https://doi.org/10.2989/1814232X.2012.725284

Shardlow, M. E. A. (2001). A review of the conservation importance of shingle habitats for invertebrates in the United Kingdom (UK). In J. R. Packman, R. E. Randall, R. S. K. Barnes & A Neal (Eds.), *Ecology and geomorphology of coastal shingle* (pp.355-377). Otley, West Yorkshire, Westbury.

Short, A. (1996). The role of wave height, period, slope, tide range and embaymentisation in beach classifications: a review. *Revista Chilena de Historia Natural 69*, 589-604. Retrieved 2020 from https://www.semanticscholar.org/paper/The-role-of-wave-height%2C-period%2C-slope%2C-tide-range-Short/a9c7692dcac1a3d4ae2a314c7b245b4ebe1f3565

Sink, K., Holness, S., Harris, L., Majiedt, P., Atkinson, L., Robinson, T., Kirkman, S., Hutchings, L., Leslie, R., Lamberth, S., Kerwath, S., von der Heyden, S., Lombard, A., Attwood, C., Branch, G., Fairweather, T., Taljaard, S., Weerts, S., Cowley, P., Awad, A., Halpern, B., Grantham, H., Wolf, T. (2012). *National biodiversity assessment 2011: Technical report. Volume 4: Marine and coastal component.* South African National Biodiversity Institute, Pretoria. Retrieved 2020, from https://www.sanbi.org/wp-content/uploads/2018/03/nba-2011-volume-4-marine-and-coastal-component-executive-summary.pdf

Snelgrove, P. V., & Butman, C. A. (1994). Animal-sediment relationships revisted: Cause versus effect. *Oceanography and Marine Biology: An Annual Review, 32,* 111-177.

Soares, A. G. (2003). *Sandy beach morphodynamics and macrobenthic communities in temperate subtropical and tropical regions - A macro ecological approach* [Unpublished doctoral thesis]. University of Cape Town.

Sousa, W.P. (1979a). Disturbance in marine intertidal boulder fields: The nonequilibrium maintenance of species diversity. *Ecology, 60*(6), 1225–1239. https://links.jstor.org/sici?sici=0012-9658%28197912%2960%3A6%3C1225%3ADIMIBF%3E2.0.CO%3B2-D

Sousa, W. P. (1979b). Experimental investigations of disturbance and ecological succession in a rocky intertidal algal community. *Ecological Monographs, 49*(3), 227–254. https://doi.org/10.2307/1942484

Sousa, W. P. (1980). The responses of a community to disturbance: The importance of successional age and species life histories. *Oecologia, 45*(1), 72-81. https://www.jstor.org/stable/4216060

Sousa, W. P. (1985). Disturbance and patch dynamics on rocky intertidal shores. In S. T. A. Pickett & P. S. White (Ed.), *The ecology of natural disturbance and patch dynamics* (pp. 101-124). Academic Press, San Diego, CA.

Stephenson, T. A. & Stephenson, A. (1972). *Life between tidemarks on rocky shores.* Freeman, San Francisco. https://doi.org/10.1002/iroh.19740590316

Takada, Y. (1992). Tide level variation of morph frequency and size structure in *Monodonta labio* (Gastropoda: Trochidae) at several boulder shores in Amakusa. *Japanese Journal of Malacologicy, 51*(3), 187-195. https://doi.org/10.18941/venusjjm.51.3_187

Takada, Y. (1995a). Inorganic content of faeces in molluscan grazers observed on a boulder shore at Amakusa. *Japanese Journal of Malacologicy, 54*(3), 195-201. https://doi.org/10.18941/venusjjm.54.3_195

Takada, Y. (1995b). Seasonal migration promoting assortative mating in *Littorina brevicula* on a boulder shore in Japan. *Hydrobiologia, 309*(1), 151-159. https://doi.org/10.1007/BF00014482

Takada, Y. (1995c). Variation of growth rate with tidal level in the gastropod *Monodonta labio* on a boulder shore. *Marine Ecology Progress Series, 117*(1-3), 103-110. https://doi.org/10.3354/meps117103

Takada, Y. (1996). Vertical migration during the life history of the intertidal gastropod *Monodonta labio* on a boulder shore. *Marine Ecology Progress Series, 130*(1), 117-123. https://doi.org/10.3354/meps130117

Takada, Y. (2001). Activity patterns of the herbivorous gastropod *Monodonta labio* on a boulder shore at Amakusa, Japan. *Venus, 60*(1-2), 15-25.

Takada, Y. (2003). Dimorphic migration, growth, and fecundity in a seasonally split population of *Littorina brevicula* (Mollusca: Gastropoda) on a boulder shore. *Population Ecology. 45*(2), 141-148. https://doi.org/10.1007/s10144-003-0151-y

Takada, Y. (2008). Contrasting characteristics in increasing and decreasing phases of the *Nerita japonica* (Mollusca: Gastropoda) population on a boulder shore. *Population Ecology, 50*(4), 391-403. https://doi.org/10.1007/s10144-008-0091-7

Tõnisson, H., Orviku, K., Kont, A., Suursaar, Ü., Jaagus, J. & Rivis, R. (2007). Gravel-pebble shores on Saaremaa Island, Estonia and the relationship with formation conditions. *Journal of Coastal Research* (50), 810-815.

Tucker, L. (2012). *Boulder shores: A unique ecosystem in need of conservation* [Honours Dissertation, Department of Biological Sciences, University of Cape Town, South Africa].

Tucker, L., Griffiths, C. L., Schroeter, F., & Vetter, H. D. (2017). Boulder shores in South Africa - A distinct but poorly documented coastal habitat type. *African Journal of Marine Science, 39*(2), 193-202. http://dx.doi.org/10.2989/1814232x.2017.1329167

Van Sickle, J. (1997). Using mean similarity dendrograms to evaluate classifications. *Journal of Agricultural, Biological, and Environmental Statistics, 2*(4), 370-388. Doi:10.2307/1400509

Warnes, G. R., Bolker, B., Bonebakker, L., Gentleman, R., Huber, W., Liaw, A., Lumley, T., Maechler, M., Magnusson, A., Moeller, S., Schwartz, M., & Venables, B. (2020). Gplots: Various R Programming Tools for Plotting Data. R package version 3.1.1. https://CRAN.R-project.org/package=gplots

Weatherington, B. (2014). *Geographical and vertical changes in the drift line isopod fauna of rocky shores in the South-Western Cape, South Africa* [Unpublished Honours Thesis]. Department of Biological Sciences, Univeristy of Cape Town, South Africa.

Wentworth, C. K. (1922). A scale of grade and class terms for clastic sediments. *The Journal of Geology, 30*(5), 377-392. Retrieved 2020, from https://www.jstor.org/stable/30063207

Wickham, H. ggplot2: Elegant Graphics for Data Analysis. Springer-Verlag New York, 2016.

Wickham, H. (2020). forcats: Tools for Working with Categorical Variables (Factors). R package version 0.5.0. https://CRAN.R-project.org/package=forcats

Wickham, H., Henry, L., Pedersen, T.L., Luciani, Matthieu Decorde, M., & Lise, V. (2020). svglite: An 'SVG' Graphics Device. R package version 1.2.3.2. https://CRAN.R-project.org/package=svglite

Wilke, C. O. (2020). ggtext: Improved Text Rendering Support for 'ggplot2'. R package version 0.1.0. https://CRAN.R-project.org/package=ggtext

Williams, J. J., Bell, P. S., Betteridge, K., & Thorne, P. (2002, July 7-12). *Transport of coarse sand by tidal currents and waves* [Paper]. Coastal Engineering 2002: Solving Coastal Conundrums - 28th International Conference, Cardiff, Wales. https://doi.org/10.1142/5165

Williams, J. J., Bell, P. S., & Thorne, P. (2003). Field measurements of flow fields and sediment transport above mobile bed forms. *Journal of Geophysical Research 108*(C4), 6-1 - 6-36. https://doi.org/10.1029/2002JC001336

Wieser, W. (1959). The effect of grain size on the distribution of small invertebrates inhabiting the beaches of Puget Sound. *Limnology and Oceanography, 4*(2), 181-194. https://doi.org/10.4319/lo.1959.4.2.0181

Wooldridge, T., Dye, A. H., & McLachlan, A. (1981). The ecology of sandy beaches in Transkei. *South African Journal of Zoology, 16*(4), 210-218. https://doi.org/10.1080/02541858.1981.11447759

WoRMS Editorial Board (2020). World Register of Marine Species. Available from http://www.marinespecies.org at VLIZ. Accessed 2020. https://doi.org/10.14284/170

 Sampled and extracted biomass (g/m²) per taxa for all 58 sites in chapter 1 with total biomass (g/m²) and species richness (#) per site.

Species/Taxa	Fine Sand (0.125-<0.25)						Medium Sand (0.25-<0.50)		Very Coarse Sand (1.00-<2.00)		Granule (2.00-<4.00)	Pebble (4.00-<64.00)							Cobble (64.00-<256.00)						Boulder (256.00+)						Rocky (N/A)					
	Geelbek	Silwerstroomstrand	Rietbaai	Bottelary	Bally Melkbosstrand	Ysterfontein	Schrywershoek	Bloubergstrand	Blouberg South 2	Blouberg North 2	Blouberg North 1	Blouberg	Blouberg South 1	Gansekraal South	Gansekraal	Grotto Bay North	Grotto Bay	Hermanus	V&A Waterfront 1	Gansekraal North	Miller's Point	Grotto Bay South	Kagelbaai	Moullie Point 1	Tucker Melkbosstrand 1	Kommetjie North 1	V&A Waterfront 2	Kommetjie South 1	Jacobsbaai South 1	Jacobsbaai North 1	Moullie Point 2	Tucker Melkbosstrand 2	Kommetjie North 2	Kommetjie South 2	Jacobsbaai South 2	Jacobsbaai North 2
chitona garnoti																							4.34	95.56					24.22	114.50	12.41					
ebhayiensis																								0.06	0.09											
inella capensis																								0.75												
rina knysnaensis	0.00		0.00	0.01																				7.44	3.09			0.44	41.84	17.09	0.91	5.09	10.13	8.22	14.66	3.66
opsis glomerata																										3.50										
a sulcatus																													25.00							
is macrophthalma																								0.06												
bioides africanus												0.06	0.03		0.06				0.06																	
ca palmata	0.05		0.00	0.00			0.01																													
sca spp.																								0.03												
holis squamata																								3.44						0.03	0.38	0.06				
ra capensis																								6.72	0.19					1.31	1.09					
oe africana																														0.22	0.03					
a achaeus																																0.06				
o lactea																															4.84					
a maritima									1.52	1.43	0.03	0.09	0.09																							
rgueleni																														0.03						
te grandicornis																								0.09	0.84			0.78	0.13		16.28	0.72	2.34	2.47		0.06
la tricolor																			5.06					0.34						1.50	10.94		0.72			
ardia spp.																								187.44							470.59	0.06				
ea globulus				0.63																																
ea isosceles			0.43																																	
ea spp.	0.09																																			
mya atra																			13.47					6.03		0.63		1.94	75.63	259.59	647.81	845.81		188.28		
sia sp.																								0.84							1.50					
glassus capensis			0.59				0.59																													
a glandula																									0.09		0.44				10.19	3.84			0.03	16.63
ta chelata																															0.13					

Species	1	2	3	4	5	6	7	8	9	10	11	12	13	14	15	16	17	18	19	20	21	22	23	24	25
Bullia digitalis	0.27		2.72	3.67																					
Bullia tenuis	0.74																								
Bunodactis reynaudi																				14.78	4.81				
Bunodosoma capense														3.72		59.38		28.28	54.69	16.66	3.06				
Burnupena cincta														19.03											
Burnupena lagenaria														14.31											
Burnupena spp.															227.72	247.63	176.47	208.84	76.53	197.03	96.97		2.69	1.50	
Calliopiella michaelseni															0.03			5.16							0.19
Callithamnion collabens																									32.38
Callithamnion spp.																							1.59	6.44	
Callochiton dentatus																				4.75					
Capeorchestia capensis			0.00	0.01			0.06		0.22	1.69	1.53	0.09	4.56												
Caprella equilibra															0.03						0.03				
Cardita variegata																0.97					0.06				
Carditopsis rugosa	0.02	0.01	0.03		0.02																				
Carpoblepharis flaccida															0.63										
Caudacanthus ustulatus															3.78		17.25			10.94	9.44		30.94		
Centroceras clavulatum															0.44			20.38							
Ceradocus rubromaculatus															0.84	0.09									
Ceramium arenarium															4.06										
Ceramium atrorubescens																								13.34	
Ceramium capense																		7.50							34.97
Ceramium planum																	0.69				1.59				
Ceramium spp.																		28.88							
Cercyon maritimus											0.03														
Cerebratulus fuscus	0.00		0.00	0.01	0.01	0.00		0.19																	
Chaetopleura papilio															21.84										
Champia lumbricalis																		123.59			34.72	44.72		37.50	55.19
Chironomidae larvae	0.00																								
Chordariaceae sp																				10.06					
Chorisochismus dentex															12.69										
Choromytilus meridionalis																				9.91					
Chthamalus dentatus																						0.25			
Cinysca dunkeri															0.75	5.59									
Cirolana spp.																				0.22					
Cirolana undulata															0.28					3.31					
Cirolana venusticauda															0.97			0.25			0.06				
Cirreformia tentaculata	0.44	0.04	0.10		0.19																				
Cladophora capensis															2.34			2.63			17.25			3.53	
Cladophora contexta																						0.25			
Clibanarius spp.															4.38			1.00							
Clinus agilis															54.88										

The table's column headers are cut off above the top of the page and the species labels are clipped at the left margin. Values are placed in their columns by horizontal position. The table is split into column blocks; the species column is repeated in each block.

Columns 1–8:

Species								
…nella sinuata								
…nella spp.								
…omenia sinuosa								
…sus algoensis								
…idula porcellana								
…opsis robusta				0.02	0.01			
…nopsis sp.		0.00						
cyanobacterial mat								
…ogropus punctatus								
…adusa filosa			0.00	0.00				
…bula granatina								
…bula miniata								
…bula oculus								
…odocella nublevis								
…elella edwardsii	0.11		0.06	0.03			0.04	
…orafismarella scutellium								
…dropoma corallinaceum								
…a formidabilis								
…o echinata								
…ichopodidae larvae	0.01		0.01	0.00				
…ichopodidae sp.								
…ax serra		1.48			32.67	107.14		3.88
…lone schultzei								
…hestia dassenensis								
…onium spp.								
…lymene sp.	0.01		0.02	0.01			0.05	
…lustoma sp								
…ce aphroditois								
…ariambus fallax								
…rostine capensis								
…dice barnardi		0.08						
…dice keraleyi		1.03						
…dice longicornis					0.81	2.64		
…rolana latipes		0.01			0.57	1.27		
…rolana natalensis		0.00			0.03	0.09		1.63
…phaeroma antikraussi								
…phaeroma brevitelson								
…phaeroma iglloecetes	0.04			0.02			0.02	
…phaeroma kraussi								
…phaeroma larviusculum								
…phaeroma planum								
…phaeroma truncatitelson								

Columns 9–16:

Species								
…nella sinuata								
…nella spp.								
…omenia sinuosa								
…sus algoensis								
…idula porcellana								
…opsis robusta								
…nopsis sp.								
cyanobacterial mat								
…ogropus punctatus								
…adusa filosa								
…bula granatina								
…bula miniata							24.88	
…bula oculus							1.81	
…odocella nublevis								
…elella edwardsii								
…orafismarella scutellium								
…dropoma corallinaceum								
…a formidabilis								
…o echinata	0.50		0.03	4.59			23.59	
…ichopodidae larvae								
…ichopodidae sp.								
…ax serra								
…lone schultzei								
…hestia dassenensis			1.56		0.66			0.13
…onium spp.								
…lymene sp.								
…lustoma sp							23.28	
…ce aphroditois						64.06		
…ariambus fallax								
…rostine capensis						1.19		0.03
…dice barnardi								
…dice keraleyi								
…dice longicornis								
…rolana latipes								
…rolana natalensis								
…phaeroma antikraussi						0.16		
…phaeroma brevitelson						0.16		
…phaeroma iglloecetes								
…phaeroma kraussi						0.34		
…phaeroma larviusculum						0.06		
…phaeroma planum						1.25		
…phaeroma truncatitelson		0.06						1.56

Columns 17–22:

Species						
…nella sinuata						0.03
…nella spp.	10.97					
…omenia sinuosa	2.66					0.44
…sus algoensis						
…idula porcellana	3.88	2.59		35.41		1.00
…opsis robusta						
…nopsis sp.						
cyanobacterial mat						
…ogropus punctatus	50.41	44.53				
…adusa filosa						
…bula granatina	1355.44	112.25		285.78		403.50
…bula miniata		0.56				
…bula oculus	1.63	239.56				0.78
…odocella nublevis	0.03					0.03
…elella edwardsii						0.03
…orafismarella scutellium						0.63
…dropoma corallinaceum						
…a formidabilis	0.63			0.31		
…o echinata	0.59					
…ichopodidae larvae						
…ichopodidae sp.	0.03					0.03
…ax serra						
…lone schultzei	45.31					
…hestia dassenensis						
…onium spp.					0.44	
…lymene sp.						
…lustoma sp						
…ce aphroditois			59.66			
…ariambus fallax	0.03				0.03	
…rostine capensis			7.88			
…dice barnardi						
…dice keraleyi						
…dice longicornis						
…rolana latipes						
…rolana natalensis						
…phaeroma antikraussi		0.63			0.28	
…phaeroma brevitelson						0.03
…phaeroma iglloecetes						
…phaeroma kraussi						
…phaeroma larviusculum	56.25		323.06			
…phaeroma planum	1.56		0.34			
…phaeroma truncatitelson						

Columns 23–29:

Species							
…nella sinuata							
…nella spp.							
…omenia sinuosa							
…sus algoensis	4.03						
…idula porcellana		14.03		1.09		0.09	
…opsis robusta							
…nopsis sp.							
cyanobacterial mat				24.53			
…ogropus punctatus	72.63		145.97				
…adusa filosa							
…bula granatina	485.44	504.59	1016.13	501.56	35.47	127.22	483.75
…bula miniata							
…bula oculus				501.47		90.63	
…odocella nublevis							
…elella edwardsii							
…orafismarella scutellium							
…dropoma corallinaceum	9.38						
…a formidabilis	4.84	0.78	0.06	0.31			0.19
…o echinata							
…ichopodidae larvae							
…ichopodidae sp.							
…ax serra							
…lone schultzei							
…hestia dassenensis							
…onium spp.							
…lymene sp.							
…lustoma sp							
…ce aphroditois							
…ariambus fallax							
…rostine capensis							
…dice barnardi							
…dice keraleyi							
…dice longicornis							
…rolana latipes							
…rolana natalensis							
…phaeroma antikraussi							
…phaeroma brevitelson							
…phaeroma iglloecetes							
…phaeroma kraussi			1.56		0.16		
…phaeroma larviusculum							
…phaeroma planum							
…phaeroma truncatitelson			0.88		0.19		

Species																										
Exosphaeroma varicolor																	6.50	0.22				1.31				
Fissurella mutabilis																	49.19	9.34	3.31		4.44	39.25	1.41	0.69	1.19	
Fucellia capensis						0.02	0.84		0.09																	
Gastrosaccus brevifissura	0.01																									
Gastrosaccus psammodytes			0.21	0.39																						
Gastrosaccus sp.	0.65	0.01			0.01																					
Gelidium pristoides																							5.91		32.94	
Gelidium spp.																				1.88					10.97	
Gibbula capensis																	0.53									
Gibbula multicolor																	0.06						0.03			
Gibbula zonata																	3.19	0.50		1.38						0.03
Gigartina polycarpa																	16.56		115.63			141.38	175.59	267.19	127.44	
Glycera tridactyla	0.01	0.03	0.04		0.01																					
Glyptidotea lichtensteini																	41.72									
Gnathia africana																				0.16						
Golfingia capensis																	0.03					13.69	0.03			
Guinusia chabrus																						17.78				
Gunnarea gaimardi																	58.81			0.69	45.41	48.00	77.22	0.91		
Halianthella annularis																			0.63							
Haliotis midae																	2.19									
Helcion dunkeri																	0.47		0.03		0.03		1.13	0.72		
Helcion pectunculus																0.34	0.28	121.13	22.88	5.66	68.84	115.81				
Helcion pruinosa														1.44		25.81	36.44	82.66	10.91	16.38	6.13	1.69	0.88	0.13	1.00	0.03
Hildenbrandia lecannellierii																		15.63								
Hymenena venosa																	2.19									
Hymenosoma orbiculare	0.02	0.01	0.03		0.01																					
Hypnea ecklonii																							5.50			
Isanthus capensis																	0.47		1.81	7.13	1.53	0.03				
Ischnochiton bergoti																0.50	0.28			0.31						
Ischnochiton oniscus																0.03				0.03						
Ischnochiton textilis																1.50			0.09							
Ischyromene australoides																		1.16								
Ischyromene huttoni																	0.47				1.41	0.31		1.50		
Ischyromene macrocephala																	0.22				3.19					
Ischyromene ovalis																			0.13		0.03					
Ischyromene scabricula																							0.06			0.19
Japygidae sp.								0.03																		
Jellyella tuberculata																			0.03							
Joeropsis stebbingi																	0.03									
Kraussilichurus kraussi			0.00																							
Kraussina rubra																	0.09									
Laminaria pallida																						191.09				

Species																									
sesia marina																						0.41			
mas johnstonei	0.01		0.02	0.78			0.30																		
onohus semirechis																7.78				25.00	4.72	1.81	1.03		
ntura laevigata							0.00																		
dilatata										0.25	0.81	0.03	60.53	135.41	1.16	0.91			21.75		0.03				
sa medusa																					0.34				
rineris coccinea																							0.31		
rineris sp.								0.00	0.03																
nasta ceratina							0.00									0.41		0.34	0.03		10.53		0.19		0.06
ce natalensis																					1.09				
a spp.																					0.03				
tbulophoxus sp.		0.01																							
niscus apandifrons														0.63				40.41							
oyna depressa	0.01			0.06			0.00																		
oyna elituera														15.19						0.19					
sella capensis														10.25				260.09		47.22					
e rosea															0.63	11.34		2.97							
a machaera														0.25											
a zeylanica	0.01																								
thura catenda														0.13											
moclinus dorsalis														145.31					38.41		9.38				
stella myositis												0.88	3.09		38.31										
sa galloprovincialis														5.53			0.38	203.06	106.69	10.88	1210.88		226.25	241.41	242.75
ris laevigata																2.50									
rtus kraussianus	0.00		0.00	0.00			0.00																		
rtus speciosus	0.00		0.00	0.00																					
lana hirtipes			0.00	0.00																					
ta capensis				0.00										0.03			0.78								
yt capensis	0.00	0.08			0.02	0.01	0.00																		
tys spp.																			1.66						
glossum binderianum					0.00	0.00												7.94							
bia sp.					0.00	0.00																			
genia erinacea															0.63				1.59				6.88	2.03	
genia ovata																								9.19	
omplana erythrotaenia															0.72										
sastus latericeus	0.06		0.00	0.05			0.07																		
lla dubia																								8.91	
la squamosa														7.53											
tostoma capense			0.00																						
della maculata																					0.63				
thrix fragilis														2.38											
ta angraequensis	0.23		0.11	0.36			0.20													4.09					

	1	2	3	4	5	6	7	8	9	10	11	12	13	14	15	16	17	18	19	20	21	22	23	24	25	26	27	28	29	30	31	32	33
Oxystele antoni																				26.72	56.38	300.41	187.88		247.91	33.09	226.38	0.13	13.16		12.88		2
Oxystele impervia																									31.88	18.66							
Oxystele tigrina																				14.78	332.16	188.19	276.50		31.66	224.50	464.34	9.97					
Pachymenia orbitosa																												104.69		76.16	134.59	127.78	12
Pachophaleria capensis						0.01																											
Pagurites gumianus																						1.94											
Palaemon pacificus						0.00																											
Paraglossum pagenstasii																									0.31								
Paramoera bidentata																	0.19																
Paramoera capensis	0.00		0.00	0.01		0.01							1.50	0.16			0.06	0.06	0.03			0.16	0.03		0.47	0.03		0.09	0.19	0.06	0.16		0
Paracridae, unidentified sp			0.00	0.01		0.00																											
Parechinus angulosus																					1293.19	52.69			146.44	1153.13	95.72						
Paridotea rubra																																	0
Paridotea ungulata			0.00	0.00																													
Partsocladius perforatus																					0.03				0.13			1.59			3.38	2.03	
Partsocladius stimpsoni																					1.41	17.47			0.59			0.06					
Parvelastra exigua																				10.03	17.59	20.03	37.03		9.91	2.06	23.53		0.19		2.34		
Pentacta doliolum																											20.47						
Perinereis namibia												13.66	9.09				2.59		0.03														
Perinereis nuntia			0.04			0.00																											
Perinereis vallata																														0.03			
Pherusa spp.																												7.44	2.44				
Philoscia sp																									0.31								
Philoscia sp.												0.03																					
Phyllymenia capensis																																	1
Piromis arenaus																						5.34											
Platydromia spongiosa																											1.44						
Plocamium maxillosum																											0.38						
Plocamium spp.																					1.50												
Porphyra capensis																					87.88	1489.41	3.03			14.38	19.84	428.63	246.56	176.53	1.19		7
Prionospio saldanha						0.00																											
Prionospio sexoculata	0.00		0.00	0.00																													
Procerodes sp.													0.06	0.06	0.03	0.13																	
Pseudharpinia excavata				0.02	0.00																												
Pseudonereis variegata																	1.44				2.88	0.09	1.88		7.50	18.44	0.66	48.25	25.13		4.38		3
Ptilohyale plumulosa													0.38	0.06				0.09															
Radsia nigrovirescens																					2.09	0.63	216.22		82.81	2.66	295.09	1.31	1.97		1.28		
Ralfsia verrucosa																					21.56									12.94	7.81		
Rhodophyllis reptans																					33.50									20.28			
Rhynsoplax polita																					5.44		8.06				51.00						
Roweia frauenfeldi																					129.69												
Sabellidae sp																														0.06			

alia scutellata																17.06				41.28							
alia stiriata																	444.06				242.81	581.28			23.03	8.63	43.50
ris squamata		1.47			5.66	27.30		0.00									0.44				0.06		0.63				
na tetraura	0.03		0.00	0.00			0.05														0.56						
stra barbara																							76.22				
stra cochlear																	72.31					225.25	/	220.94	398.34		727.72
stra granularis											11.50	9.78	60.75		73.44	81.03		3.81	94.13	55.09	61.81	378.31		277.75
stra longicosta													0.16				229.97										
capensis					0.65	1.12																					
tia erythroensis	0.13		0.07	0.13			0.05																				
ria capensis																						7.72					
ria compressa	0.01		0.00	0.00																							
ria concinna											0.53	0.31	7.28						52.63	0.97	12.19	12.19					
ria serrata														0.03					3.44								
mene microtylotos												5.41															
is spp.											0.03	0.47	4.22						16.63		1.69						
dium rugosum													10.97					0.44	115.47		156.25	140.16	106.38	19.25
es yendoi													0.78	0.13	16.13	0.91				46.16	50.72	92.56	63.06			25.00
iridae sp		0.00									0.03																
capensis																							0.03				
myidae sp									0.03		0.13																
ladia camptoclada																	12.78										76.16
sp										0.03	0.13										0.22	0.16					
dae sp								1.09																			
estia australis		0.07																									
estia sp									0.37																		
um brevipes											0.66					1.66			0.06								
lla tenebrosa																									1.71		
a sp.			0.02	0.00			0.00																				
a trigona	0.05		0.03	0.00			0.07																				
geton minor													0.03						0.03				0.03				
rpus capensis	0.07		0.05	0.09			0.00																				
dias capensis													3.47			1.69	0.03										
ta serrata															0.28		1.19		37.56	33.09							
concamerata													1.94														
a spp.																					2.19	0.16					
des natalensis																	3.75										
capensis													0.31		0.06						0.22						
kochii													0.06														
nerstina																						0.34	0.19				
spp.													0.03														
rium cerebriforms																	9.13					31.50					

	1	2	3	4	5	6	7	8	9	10	11	12	13	14	15	16	17	18	19	20	21	22	23	24	25	26	27	28	29	30	31	32	33	34	35
Turbo cidaris																								406.25	4.69										
Turbonilla krausai	0.00			0.00																															
Turritella capensis	0.00		0.00	0.00			0.00																												
Tylos granulatus					1.28	0.04																													
Ulva rigida																											0.88								
Ulva sp																								50.44	23.34	15.25				0.75	68.91	48.72	71.75	40.19	
Upogebia africana	0.00		0.00	0.05			0.01																												
Urothoe grimaldii	0.00		0.01	0.06	0.00	0.00	0.01																												
Urothoe sp.		0.00																																	
Urothoe tumorosa		0.00																																	
Volvarina capensis	0.00		0.00	0.00			0.00																												
Waternipora subtorquata																										9.38									
Zeuxoides helleri																								0.03		0.53		0.06						0.31	
TOTAL	1.41	5.90	1.55	2.57	44.68	143.69	1.72	5.89	1.54	2.84	0.09	0.34	0.50	17.63	15.25	3.53	3.47	62.88	0.00	24.34	226.41	5.75	158.63	4950.59	3153.91	1574.56	1.75	2701.53	2838.84	3139.22	3625.97	5214.28	9889.13	1792.19	1900.19
Species Richness	33	17	34	38	16	16	32	6	2	5	3	3	6	6	8	5	4	5	0	9	9	3	14	100	48	42	3	54	37	60	79	50	30	43	17

*Day, Brown, Harris and Nel had no biomass taken

Appendix B Sampled and extracted density (#/m²) per taxa for all 58 sites in chapter 1 with total density (#/m²) and species richness (#) per site.

Sites — Shore Type with Particle Size Range (mm)

Density (#/m²) per Site — Species/Taxa

Fine Sand (0.125–, 0.25)

Species/Taxa	Langebaan Village	Strand	Groβbek	Silverstroomstrand	Onrusval (Langebaan)	Harris Melkbostrand	Cotton	Macassar	Rietbaai	Hout Bay Harbour	Buitalay
…nicola …ti										F / C	
…nicola sp.											
…hochitona …i											
…a …dentata											
…a ebbegiensis											
…rchestia …spinosa	F / C				P						
…minella …xis											
…torina …ensis			0							0	8
…phila …ata					P						
…ohenia pluma											
…iopsis …rata											
…idium …an											
…a ovalis	P										
…a …ulcata											
…llia …phthalma											
…robioides …rus											
…sea diadema					P						
…uca palmata			22	C							
…uca spp.											
…balanus …rite											
…bicstrum …y											

Medium Sand (0.25–, 0.50) · Coarse Sand (0.50–, 1.00) · Very Coarse Sand (1.00–, 2.00) · Granule (2.00–, 4.00)

Species/Taxa	Voterfanteln Central	Hout Bay North	Harris Melkbostrand	Voterfanteln North	Voterfanteln South	Hout Bay East	Schrywershoek	Bloubergstrand	Nel Melkenberg	Harris Hermanus	Milnerton	Brown Melkenberg	Llandudno	Lynch Point 1	Blauberg South 2	Blauberg North 2
…nicola …ti																
…nicola sp.										1						
…rchestia …spinosa	P	28	2		P		4	C	5	26	A		1	1	A	A
…robioides …rus														12.50	3.13	
…uca palmata			1		0		2									

Pebble (4.00–, 64.00) · Cobble (64.00–, 256.00)

Species/Taxa	Blauberg North 1	Blauberg	Blauberg South 1	Gansbaai South	Gansbaai	Grotto Bay North	Grotto Bay	Hermanus	V&A Waterfront 1	Gansbaai North	Miller's Point	Grotto Bay South	Kogelbaai	Maullie Point 1
…hochitona …i														6.25
…a ebbegiensis														3.13
…rchestia …spinosa	C	L / C												
…minella …xis														3.13
…torina …ensis														121.88
…llia …phthalma														6.25
…robioides …rus						3.13							12.50	
…uca spp.														3.13

Boulder (256.00+) · Rocky (N/A)

Species/Taxa	Tucker Melkbostrand 1	Kommetjie North 1	V&A Waterfront 2	Kommetjie South 1	Jacobsbaai South 1	Jacobsbaai North 1	Maullie Point 2	Tucker Melkbostrand 2	Kommetjie North 2	Kommetjie South 2	Jacobsbaai South 2	Jacobsbaai North 2	Lynch Point 2	Exposed Shapes Is.	Sheltered Shapes Is.
…hochitona …i		25.00				3.13	25.00	12.50					PC	PC	C
…a …dentata															P
…a ebbegiensis	6.25												P	C	C
…minella …xis															P
…torina …ensis	65.63		9.38	84.38	112.50	31.25	131.25	5.00	368.75	156.25		62.50	A	A	L / C
…phila …ata															?
…ohenia pluma													P	P	P
…iopsis …rata															P
…idium …an													P	P	P
…a …ulcata							3.13								
…llia …phthalma													P	P	
…robioides …rus															P
…sea diadema															P
…balanus …rite													P	P	PC
…bicstrum …y															P

The table is a species occurrence/abundance matrix; column headings do not appear on this page. Columns are reproduced by their left-to-right grid position and the table is split into a left group and a right group, with the species column repeated.

Species																	
Amphipholis squamata			P														
Amphiura capensis																	
Amphitoe africana																	
Amphitoe ramondi																	
Anachis braueri																	
Anomia achaeus																	
Antholobus stimpsoni																	
Antinoe lactea	P		FC														
Anurida maritima													562.50	531.25	59.38	96.88	18.75
Aora kerguelensi																	
Aora typica																	
Apohyale grandicornis																	
Apohyale hirtipalma																	
Arabella iricolor																	
Arenicola loveni	P		LC		C		P				FC						
Areopagorites eseyops																	
Argobuccinum pustulosum	P																
Aristias symbiotica																	
Arthrocardia flabellata																	
Arthrocardia spp.																	
Assiminea globulus			A			1880											
Assiminea lanceae				125													
Assiminea spp.		116															
Aulacomya atra																	
Aulactinia sp.																	
Austromegabalanus cylindricus																	
Autolytus tuberculatus																	
Balanoglossus capensis			LC	11						8							
Balanus glandula																	
Bathyporeia cuvieri								4									
Bathyporeia gracilis					P		C		FC		P	FC					
Beania inermis																	
Berthellina granulata			P														
Betaeus jucundus			P														
Bledius sp.			13														
Boccardia cheata																	

Species																
Amphipholis squamata		165.63						6.25	31.25	9.38					PC	P
Amphiura capensis		71.88	3.13					9.38	25.00						P	P
Amphitoe africana								3.13	3.13							
Amphitoe ramondi																
Anachis braueri															PC	
Anomia achaeus										3.13						
Antholobus stimpsoni															C	C
Antinoe lactea										9.38					P	P
Anurida maritima															LC	LC
Aora kerguelensi								6.25								
Aora typica															P	
Apohyale grandicornis		6.25	15.63			68.75	3.13		837.50	53.13	193.13	256.25		9.38		
Apohyale hirtipalma															LC	LC
Arabella iricolor	6.25	3.13						12.50	12.50		25.00				P	
Arenicola loveni																
Areopagorites eseyops															P	
Argobuccinum pustulosum															P	PC
Aristias symbiotica															P	
Arthrocardia flabellata																
Arthrocardia spp.									6.25	3.13						
Assiminea globulus																
Assiminea lanceae																
Assiminea spp.																
Aulacomya atra		18.75		3.13		6.25	5.00	103.13	115.63	196.38		93.75			A	FC
Aulactinia sp.		3.13							9.38							
Austromegabalanus cylindricus															C	C
Autolytus tuberculatus																
Balanoglossus capensis																
Balanus glandula			6.25		6.25				240.63	168.75		3.13	18.75			
Bathyporeia cuvieri																
Bathyporeia gracilis																
Beania inermis																
Berthellina granulata																
Betaeus jucundus																
Bledius sp.																
Boccardia cheata									3.13							

Taxon (label truncated)	1	2	3	4	5	6	7	8	9	10	11	12	13
…ider / …erum													P
…omoldone													P
…oma / …w												P	P
…oma / …a											P	FC	FC
…ralathus													P
…neritina		P											
…gitalis	F/C	2	1 0	5 F/C	1 A 1	F/C P F/C							
…evissima	F/C			C		V							
…hodostoma		2	8 1			A							
…realis		2											
…ctia / g													
…soma									6.25 25.00		C	A	C
…ena / …cta								3.13 6.25	3.13 3.13 9.38 21.88		C	FC	C
…ena cincta											C	C	C
…eus / …s								3.13			C	FC	FC
…res / …ra								3.13					
…res / …ea											P	P	P
…ena spp.								109.38 115.63 190.63	56.25 25.00 234.38 106.25	21.88 3.13 3.13			
…na africana		P										P	P
…sier												P	P
…obius / …		P											
…obius / …		P											
…ella / …eri								3.13	337.50	12.50		P	P
…anton / …										3.13			
…anton spp.									9.38				
…ton									25.00				
…tria / …											P	P	
…nchus													P
…hestia		1		1			3.13 6.25 34.38 153.13 18.75 334.38						
…a capitata			2										
…a cicur												P	P
…sii												P	P
…a equilibra								3.13	3.13			P	P
…a penuatis											P		
…a scaara											P	P	P
…na / …tis											P		
…na spinigr											P		P

Left-hand data columns (stations 1–16):

Species	1	2	3	4	5	6	7	8	9	10	11	12	13	14	15	16
Carditopsis rugosa		3	C		1		10					3				
Ceramium capense			C													
Ceramium planum			P													
Cerebratula fuscus	F/C	0	P	0			F/C	1	1	F/C	2	F/C	1	C	C	P
Chironomidae larvae		2														
Choromytilus meridionalis	L/C		P													
Cirriformia tentaculata		16	P			2	V/C	3					12			

Right-hand data columns (percentages) and biotope codes:

Species	R1	R2	R3	R4	R5	R6	R7	R8	R9	R10	Code	
Cardita variegata				6.25				6.25				
Carditopsis rugosa											P	
Caulacanthus ustulatus							15.63		93.75		P	
Caulacanthus ustulatus												F
Celleporella hyalina												
Centroceras clavulatum					3.13							
Ceradocus rubromaculatus			15.63	3.13							C	F
Ceramium capense												
Ceramium planum							3.13					
Cereyon maritima		3.13										
Cerebratula aerugatus											P	
Cerebratula fuscus	3.13										FC	
Chaetopleura papilio			3.13								P	
Champia compressa											L/C	
Champia lumbricalis					9.38		15.63	15.63		3.13	L/C	F
Chaperia acanthina											P	
Chironomidae larvae												
Chondria capensis												
Chondrochelia navignyi											FC	
Chorinochismus dentes			9.38									
Choromytilus meridionalis						6.25					A	
Chthamalus dentatus								3.13			P	
Cinqsca dunkeri			15.63	65.63								
Ciocalypta alleni												
Circularius wilsoni												
Cirolana parva											P	
Cirolana rugicauda												
Cirolana sakianhae												
Cirolana spp.						3.13						
Cirolana sulcata											P	
Cirolana undulata			3.13			15.63						
Cirolana venusticauda			9.38		3.13	6.25					FC	
Cirriformia sp.											L/C	
Cirriformia tentaculata											P	
Cladophora capensis			3.13			6.25						

…ra																									FC
…ur																									P
…u spp.											15.63								3.13						
…lia											6.25														
…oides																								P	P
…nar																								P	
…sinuata			A												3.13								P		P
…spp.											6.25														
…agile																								C	P
…ia																									P
…oenis																3.13									
…zambicus																							P	P	P
…myota																									P
…r																							LC	FC	LC
…a	P		P							18.75	28.13	78.13		3.13		25.00	6.25		6.25			C	C	C
…ella																							P	P	P
…niopsis																							P	P	P
…sp.			1																						
…a			P																						
…nar											15.63	28.13				68.75	93.75						C	C	C
…n filosa			P	0	0																				
…xa																							P	P	P
…a										62.50	31.25	103.13		81.25	18.75	18.75	56.25	12.50	6.25	9.38	12.50	C	A	C
…miniata									3.13		3.13											P	P	P
…oculus									3.13	12.50	3.13				3.13			9.38	6.25			LC	C	C
…xa																									P
…valida			P																						
…la																								P	
…la											6.25								3.13						P
…		3	FC	2	3	2																			
…char																						P		
…aurella	P															3.13						FC	FC	C
…rum																187.50							LA	LA
…idabilis									6.25		3.13			25.00	3.13	3.13	6.25			3.13		P	P	P
…nata							6.25	34.38	65.63			175.00		9.38										
…r																									P

Species	1	2	3	4	5	6	7	8	9	10	11	12	13	14	15	16	17	18	19	20
Diogenes brevirostris	P		A																	
Diopio magna							P	P				P				P	P	P		
Dodecaceria capensis																				
Dodecaceria pulchra																				
Dolichopodidae larvae		1					1	1												
Dolichopodidae sp.																				
Donax serra	P	0	7	2	0	1	C	8	C	3	15		C	90	1	F/C	C	L/C		
Donax sordidus						0									0					
Drillia sp.																				
Dromidia hirsutissima																				
Batonella ufrostgra																				
Echinocardium cordatum																P				
Ecklonia maxima																				
Elasmopus rapax																				
Electra pilosa																				
Eledone schultzei																				
Emplectonema ophiocephalum																				
Orchestia skaawensis																				
Orchestia rectipalma																1				
Epitonium spp.																				
Erichthonius brasiliensis																				
Euclymene luederitziana			A																	
Euclymene sp.		0					1	1					3							
Eudistoma ilotum			P																	
Eudistoma sp																				
Eulalia cf. *trilineata*																				
Eunice aphroditois																				
Euparianthura falla																				
Euphrosine capensis																				
Eurydice barnardi		44/5	86/8	86/9	28	88		70												
Eurydice kensleyi		1	32/6		1	2		10			13/4	3			3					
Eurydice longicornis	P		P						P			A				C	C	C		
Eurydice sp.				21/9						7					1					
Exciroloma latipes	F/C		2	P		10	3	P			11	P	1	9		P	P	P		
Excirolana natalensis		3	1	2							11		2	8	12/7			86	2	39
Exogone verugera																				

Species	a	b	c	d	e	f	g	h	i	Code
Dodecaceria capensis										P
Dodecaceria pulchra										L/C
Dolichopodidae sp.				3.13					6.25	
Donax serra										L/C
Batonella ufrostgra										P
Ecklonia maxima										A
Elasmopus rapax										FC
Electra pilosa										P
Eledone schultzei				3.13						
Emplectonema ophiocephalum										P
Orchestia skaawensis	278.13	78.13	6.25							
Epitonium spp.								3.13		
Euclymene luederitziana										P
Eudistoma ilotum										P
Eudistoma sp					12.50					
Eulalia cf. *trilineata*										P
Eunice aphroditois				3.13			9.38			PC
Euparianthura falla						3.13			6.25	
Euphrosine capensis				3.13	3.13			15.63		
Exogone verugera										P

...vaeroma raasi																													6.25						21.88		6.25									P
...vaeroma elson																													3.13										3.13							
...vaeroma cetes		4		P							2						3																													
...vaeroma ni																													6.25												21.88	3.13				P	P
...vaeroma aculum																													6.25		175.00				11137.50											
...vaeroma son																																													P	
...vaeroma m																													12.50		12.50				6.25										P	
...vaeroma ctus																																												P		
...vaeroma atirison									A			P	1								P							15.63			40.63									28.13		3.13			
...vaeroma olor																													106.25	6.25							15.63									
...ella ulis				FC																									143.75	15.63		87.50					6.25	137.50	62.50	15.63	78.13			P	P	P
...lligera																																													P	P
...roncyphus agonus																																												P		
...lia capensis	L C																							L C	3.13	112.50	12.50																	P		
...ocelliferus					P																																							P		
...maropsis esi																																												P		
...osaceus lunea				0																																										
...osaceus nodytes		6				10	8	10				17	P	8	6	C		8	C	C	VC	FC																								
...osaceus sp.				46							2						1																													
...ya humilis																																														
...um pristoides																																							9.38		9.38						
...da becheri																																													P	
...da capensis																													21.88																		P
...da cicer																																												P		
...da multicolor																													3.13										3.13								
...da zonata																													75.00	3.13						9.38						9.38		P	P	P
...dna erpa																																					9.38	21.88	9.38								
...tina scabiosa																																														C
...atus licaudatus																																												P		
...ra sp.		2														1																														
...ra tridactyla	P			0	FC				1	VC	2		C			C	1				FC	C	P																							
...doira asteini																													162.50																	P
...ia africana																																		125.00												
...olana bularis																																												FC		
...gia capensis																													9.38								37.50	3.13						FC	FC	C
...tariopsis ssima	P				A																																									

Griffithsius latipes						2	2															
Guinusia chabrus																3.13						P	P
Gunnarea gaimardi										65.63				3.13	15.63	34.38	287.50	21.88				L/A	FC
Halianthella annularis													3.13									
Haliclona anonyma			P																			
Haliotis midae										3.13													P
Harmonia typica			P																			
Harmothoe saldanha			P																			
Harmothoe waahli			P																			
Helcion dunkeri										9.38			3.13	3.13		25.00	15.63						P
Helcion pectunculus										3.13	3.13	175.00	43.75	15.63	53.13	84.38						C	FC
Helcion pruinosus									9.38	65.63	109.38	140.63	28.13	37.50	6.25	15.63	28.13	6.25	25.00	3.13			P
Hemilepida erythrotaenia																							P
Hemiocnus insolens	P																					C	FC
Henricia ornata																						FC	P
Hippolyte kraussiana			P																			
Hippomedon inconctus			P																			
Homalina trilatera	F/C		FC																			
Hourstonius pusilla																							P
Hydractinia altispina																							P
Hymenosoma orbiculare	P		C																			
Hymenena venosa															3.13								
Hymeniacedon perlevis																						A	A
Hymenosoma orbiculare		2		1	2			1														
Hyperetione foliosa	P		P																			
Hypnea spicifera																						P	P
Isis pubescens																						P	P
Janiropsis palpalis																						P	P
Isthrippa capensis																						P	P
Ixantha capensis																15.63	3.13	12.50	25.00	3.13			P
Ixara aensamosa																							P
Ischnochiton bergoti															6.25	9.38		6.25					P
Ischnochiton oniscus															3.13			3.13					
Ischnochiton textilis															12.50				3.13				
Ischyromena buttoni																						C	P

…omene …lis																												LC	
…mene …oides																			12.50										
…omene																		9.38					43.75	15.63	43.75				
…mene …ephala																		3.13					143.75						
…omene																				6.25			3.13				P		P
…mene …ula																								3.13	12.50				
…ois …enis																											P		
…idae sp.											3.13																		
…lcata																											P		
…lendi																											FC		
…la …ulata																						3.13							
…ta stebbingi																		3.13											
…pnaeria …a																											P	P	P
…tlichina	P	A		C	0	P		0	P																				
…na rubra																		3.13											
…sina sp.																													P
…ria pallida																											C	FC	P
…a gardineri																											FC	FC	
…maropsis …a																											FC		
…mae …toni	?	1	FC		2	40	28																						
…otus senile																		12.50				3.13	6.25	15.63	12.50		PC	FC	PC
…dara …na							0																						
…elia …tra																											P		
…elia timida																											P		
…oe …di		P																											
…latata												40.63	37.50	6.25	1234.38	5390.63	137.50	59.38			1256.25		3.13				C	C	C
…sa …lata																													P
…ruber																											P		
…lobus …dis		P																					3.13						
…medusa																											P	P	P
…ellaria …ria																													
…neria …sa																								6.25			FC		P
…neria sp.			0				0			6.25	62.50																		
…assa …a	P	C					0											53.13		31.25			3.13	6.25	421.88	15.63	P	P	P
…e natalensis																									9.38		FC	P	P
…phrys …a		LC																											

Species	1	2	3	4	5	6	7	8	9	10	11	12	13	14	15	16	17	18	19	20	21	22	23	24	25	26	27	28	29	30	31	32	33	34	35	36	37	38	39	40	41	42	43
Mactra glabrata					P																																						
Macra hirondellei																																											P
Macra spp.																																			3.13								P
Mandibulophausa sp.				1																																							
Marionina spatulifrons																										3.13	45.00																
Marphysa capensis																																										FC	P
Marphysa depressa	L/C		0		C							1						0																									
Marphysa elituri																													6.25						3.13								
Marphysa variegnea																																											
Maxmuella capensis																																3.13					6.25						
Melibe rosea																															3.13	9.38	3.13										
Melita machaera																													9.38														
Melita orgasmos																																											P
Melita zeylanica			4																																								
Melliarya floridensata																																											
Menipea triserrata																																											
Mesanthura catenula																													6.25														P
Msevella tulipa					P																																						
Molgula arenata																																											
Mureus concaviform																																											P
Mureanochinus dorsalis																													25.00		15.63		6.25										
Mytasteila reyssnieri																								65.63	131.25			1625.00															
Mytilus galloprovincialis																													9.38		3.13	53.13		15.63		21.88		340.63	140.63	143.75	131.25		
Naineris laevigata																														3.13												FC	P
Nassarius kraussianus			1							0		1					1																										
Nassarius speciosus			2		P					0		0																															
Natatolana hirtipes					P					0		0																															
Nautilocorystes ocellata					P																																						
Neanthes operta																																											P
Nebalia capensis					P						1																		3.13	25.00													P
Nephtys capensis		P	3	1	1	PC	1	6	5		P		12	P	4	0		C	0	5	C	V/C	P																				
Nephtys hombergi					P						F/C			P																													
Nephtys spp.																																3.13											
Nereiphylia castanea																																											P
Neurogblastum londonianum																														6.25													
Namibia capensis							0																																				

...chia	L C																							P	P
		P																						P	P
...ia																		3.13			3.13		L C	L C	L C
...ia																					25.00				P
...iana ...ia																6.25								C	FC
...ir	26	C		3	6				11																
...balanus																							A	C	P
...bia																					3.13		C	FC	P
...a																3.13							FC	P	P
...i																									P
...ir		L C																							
...siculata																								P	
...oma			0																				P		P
...vulgaris																							A	C	
...lis																							P	P	P
	P																						P	P	
...a																			18.75						
...a fragilis		P														9.38							FC	FC	FC
...ensis	P	8	A	3	6		6												25.00						
			0																						
...sp.		1				2																			
...erstonei																							P	P	
...nctatus	P																								
...atoni										40.63	90.63	565.63	118.75		165.63	31.25	212.50	6.25	25.00		12.50		9.38	A	A	A
...mpervia																56.25	12.50									
...grina										3.13	65.63	62.50	38.13		3.13	21.88	87.50	6.25						C	C	C
...ta																					3.13		6.25	C	C	C
...eria			0																						
																							P		P
...pacificus		P			0																			FC		P
...a																							C	C	P
...a					34.38																					

Left-hand (count / presence) columns:

Taxon	C1	C2	C3	C4	C5	C6	C7
Paramoera capensis	2	P		0	3	4	
Paranthura punctata							
Paraonidae, unidentified sp				1	3	2	
Paraphoxus oculatus		P					
Parapseudes spongicola							
Parechinus angulosus	P	P					
Paridotea rubra							
Paridotea ungulata	P	C		0	0		
Parisocladus perforatus							
Parisocladus stimpsoni							
Parvakastra exigua	P	FC					
Pavoclinus pavo							
Pectinaria capensis		P					
Pegusa nasuta	P	P					
Pentacta doliolum							
Perinereis falsovariegata							
Perinereis namibia							
Perinereis nuntia					1	0	
Perinereis nutia		P					
Perinereis vallata							
Perioculodes longimanus		P	2	1		10	
Perna perna							
Petaloproctus terricolus		P					
Phascolosoma agassizii							
Pherusa monroi							
Pherusa spp.							
Philonthus sp.							FC
Philoscia sp							
Philoscia sp.							
Phoxocephalidae sp	20		1	15		2	
Phoxostoma variegatus							
Phyllochaetopterus socialis							
Phyllymenia capensis							
Phylo foetida	P	P					
Pilumnoides perlatus							
Pilumnus minutus							

Right-hand (numeric) columns and final code column:

Taxon	N1	N2	N3	N4	N5	N6	N7	N8	N9	N10	N11	N12	N13	N14	N15	N16	N17	N18	N19	N20	N21	Code
Paramoera capensis		390.63	28.13				6.25	15.63	3.13		12.50	3.13		40.63	3.13		9.38	18.75	6.25	31.25	3.13	C
Paranthura punctata																						FC
Paraonidae, unidentified sp																						
Paraphoxus oculatus																						
Parapseudes spongicola																						P
Parechinus angulosus										168.75	15.63			3.13	46.88	3.13						C
Paridotea rubra																					3.13	
Paridotea ungulata																						
Parisocladus perforatus										3.13				6.25			37.50		98.75	21.88		P
Parisocladus stimpsoni									9.38	190.63			3.13			3.13						
Parvakastra exigua									25.00	31.25	53.13	62.50		12.50	6.25	40.63		3.13		9.38		C
Pavoclinus pavo																						
Pectinaria capensis																						
Pegusa nasuta																						
Pentacta doliolum																6.25						C
Perinereis falsovariegata																						P
Perinereis namibia		2156.25	1459.38				93.75		6.25													
Perinereis nuntia																						
Perinereis nutia																						
Perinereis vallata																		3.15				
Perioculodes longimanus																						
Perna perna																						P
Petaloproctus terricolus																						
Phascolosoma agassizii																						P
Pherusa monroi																						
Pherusa spp.													28.13	9.38								
Philonthus sp.																						
Philoscia sp												6.25										
Philoscia sp.	18.75																					
Phoxocephalidae sp																						
Phoxostoma variegatus																						P
Phyllochaetopterus socialis																						P
Phyllymenia capensis																					3.13	
Phylo foetida																						
Pilumnoides perlatus																						C
Pilumnus minutus																						P

mis arenarius																										3.13												P		P
dromia dosa																															3.13									
ereis dis																																						P		
ereis rilii																																						FC	P	P
nium lonum																															3.13									
daria fera																																							P	
daria a																																						P	P	P
cerus crisatus																																						P		
carus spicuus																																						P		
cera nsis		P																																						P
heria atolli																																						LC	FC	LC
irrus etis																																						P		
fora ciliata																																						LC		
astia lius																																						P		P
oe endrina																																							P	
yra capensis																									21.88	6.25					6.25	3.13	18.75	12.50	12.50	6.25	6.25	LA	LA	LA
ospio hia									0																															
ospio lata	1	P		1	0																																			
rastea odes																																						LC		
rodes sp.															81.25	9.38	3.13	71.88																						
cirrineris derma																																						P	P	P
hyale ha																																						P	P	P
nogobius ensis		C																																						
actinia nda																																						FC	C	C
omegamphop opsis		LA																																				P		
onereis ata																					3.13				37.50	3.13	3.13		87.50	21.88	3.13	378.13	175.00		53.13		9.38	P		
opotamilla rmis																																						LC		
iphonia hyila																																								C
yale losa																	81.25	9.38				9.38																		
stolonifera																																						A	C	C
rimaera ipes																																						P		
na amis																																						P		
a drescens																									18.75	3.13	484.38		184.38	28.13	237.50	9.38	15.63		25.00			P	FC	C
a verrucosa																									15.63									9.38						

Species																	
Rhyssoplax polita																	
Roweia frauenfeldi																	
Sabellidae sp																	
Sarcothalia radula																	
Sarcothalia scutellata																	
Sarcothalia stiriata																	
Schistomeringos neglecta			P															
Schizoporella sp.																	
Scissodesma spengleri										4						VC		
Scolelepis squamata	P	3	72	84	5	12	FC	909	FC	107	1	VC	0	1	C	A	P
Scoletoma tetraura	FC	17	FC				0	2		2			25		7		
Scoloplos sp.		16			20	48								6			
Scutellastra argenvillei																	
Scutellastra barbara																	
Scutellastra cochlear																	
Scutellastra granularis																	
Scutellastra longicosta																	
Sepia typica			P															
Sertularella africana																	
Sigalion capensis				1						2							
Simplisetia erythraeensis		9	P				6	10					7				
Sinelobus stanfordi																	
Siphonaria capensis																	
Siphonaria compressa		2					0	1									
Siphonaria concinna																	
Siphonaria costata																	
Siphonaria serrata																	
Siphonenteron bilineatum																	
Smittina sp.																	
Sphaeramene microtylotos																	
Spio filicornis	LA																	
Spiroplax spiralis			P															
Spirorbis spp.																	
Spirbrachidium rugosum																	
Spongites yendei																	

Species														code	code
Rhyssoplax polita		3.13	3.13		3.13										P
Roweia frauenfeldi		3.13												LA	P
Sabellidae sp									3.13						
Sarcothalia radula														FC	LA
Sarcothalia scutellata							3.13								
Sarcothalia stiriata					15.63				12.50	25.00			3.13		
Schistomeringos neglecta															
Schizoporella sp.															
Scissodesma spengleri															
Scolelepis squamata					6.25				6.25		65.63				
Scoletoma tetraura									6.25					P	
Scoloplos sp.															
Scutellastra argenvillei														A	P
Scutellastra barbara											3.13			C	FC
Scutellastra cochlear					9.38					65.63		37.50	93.75	LA	FC
Scutellastra granularis			9.38	9.38	18.75	15.63	25.00	15.63	21.88	21.88	3.13	86.38	68.75	A	A
Scutellastra longicosta	3.13				6.25										
Sepia typica															
Sertularella africana															P
Sigalion capensis															
Simplisetia erythraeensis															P
Sinelobus stanfordi															P
Siphonaria capensis									12.50					P	P
Siphonaria compressa															
Siphonaria concinna		3.13	3.13		12.50				25.00	6.25	15.63	9.38			
Siphonaria costata															P
Siphonaria serrata		3.13							3.13					FC	A
Siphonenteron bilineatum															P
Smittina sp.															
Sphaeramene microtylotos				137.50											
Spio filicornis															
Spiroplax spiralis															
Spirorbis spp.		71.88	78.13		937.50				1562.50		375.00				
Spirbrachidium rugosum									9.38				3.13	LC	C
Spongites yendei		3.13	3.13						3.13	6.25	3.13	3.13	3.13		

…ylixidae sp		2							6.25													
…s cespexis																		3.13			P	P
…num nonus																					P		
…ecrylidae sp										3.13	18.75											
…ciiniia vicielii																				3.13			
…is cocnopn																							P
…ne sp												6.25					21.88	3.13				
…ermillaris																					P	P	
…corteposta																					C	P	
…theese mensa			P																	P	P	P
…tidne sp											3.13											
…faesiia li		10	A																			
…beuiks rdipes			P																			
…iheutia sp							33															
…iheutia sp								19														
…ylum v													12.50			6.25	6.25			P	P	P
…ratira teria			FC																	P		
…ni sp				4	0		1															
…sa trigoa	15			7	1		22															
…idae sp						2																
…agraos													3.13		3.13			3.13				
…rlepur is	1		L C	0	2		0															
…oldita li													159.38			84.38	3.13			P		
…tita sorta																					P	P
…tita sdari																				P	P	
…tita serrata														3.13		3.13	9.38	9.38		C	A	A
…romos rputum																				P		
…ita serrata													9.38							P	C	A
…ar cenus																				P	P	P
…ne spp																	3.13	6.25				
…ne minor			P																			
…ria losta																				P		
…sodes li															78.13					P	P	C
…aurva	P		P																	C	FC	FC
…ta rotia																				P		
…n cispenta													46.88	3.13			12.50			P	P	P
…kochii													9.38									
…neritina																	25.00	31.25		P	P	P

Tricolia spp.																																									3.13														
Trididemnum cerebriforme																																														9.38			6.25						
Tubularia sp																																																						C	
Turbicellepora avicularis																																																							
Turbo cidaris																																									25.00	3.13													
Turbonilla krausi			0								0																																											P	
Turritella capensis			1		L A				1		2							0																																					
Tylos capensis																							P																																
Tylos granulata	L C						0					13			39		F C					P			L C																														
Ulva capensis																																																						L A	L A
Ulva intestinalis					FC																																																		
Ulva rigida																																												3.1 3											
Ulva sp																																									3.13	87.50	3.13				3.13	21.88	5.00	3.13	6.25		12.50		
Upogebia africana			0						0		1							1																																					
Upogebia capensis																																																							
Urothoë elegans					P					P			P				P					P	P	P																															
Urothoe grimaldii			1		FC				4		19			1				5																																					
Urothoë pinnata		75					29	16												1																																			
Urothoe pulchella					L C																																																		
Urothoë sp.				0																																																			
Urothoë tumorosa				0																																																			
Vaughtia scrobiculata																																																							
Venerupis corrugata	P				L C																																																	FC	
Venus verrucosa					P																																																		
Virgularia schultzei	C																																																						
Volvarina capensis	P		1		FC				2		2							1																																					
Volvarina sp																																																							F
Watersipora subtorquata																																											3.13												
Zeuxoides helleri																																									3.13		137.5 0		15.63					62.50					
Zygamea warreni																																																						P	
TOTAL	0	576	263	1343	0	1235	121	202	184	0	2017	1071	0	148	223	156	0	151	210	41	70	0	0	0	0	566	659	125	59	216	2906	1728	222	44	1488	0	178	5673	475	1853	2803	1906	1958	19	15103	680	1844	3384	3139	789	1572	553	580	0	0
Species Richness	44	13	33	17	10 2	15	16	14	34	17	38	13	17	16	12	3	14	32	6	16	7	15	20	11	6	2	5	3	3	6	6	8	5	4	5	0	9	9	3	14	88	46	33	3	50	35	58	74	49	28	29	11	21	21 6	1 6

*Bally had no abundance taken.
*Brown used the following abbreviations: P = Present, FC = Fairly Common, C = Common, VC = Very Common, and A = Abundant
*Day used the following abbreviations: A = Abundant, LA = Locally Abundant, C = Common, LC = Locally Common, FC = Fairly Common, and P = Present

Appendix C Species presence per taxa across each of the 12 sites sampled for Chapter 2 and totaled per site. Calculated as total occurrence (#) and percent occurrence (%) per taxa.

	Species Richness (#) per Site Total		Sites												Total Occurrence (#)	Total % Occurrence
	Sites		Blouberg	Miller's Point	V&A Waterfront 1	Kogelbaai	Hermanus	Grotto Bay South	Blouberg South	Grotto Bay North	Grotto Bay	Ganzekraal North	Ganzekraal	Ganzekraal South		
	Sample date		4/7/2019	4/8/2019	5/17/2019	5/17/2019	6/17/2019	08/13/2019	09/13/2019	09/16/2019	09/16/2019	09/30/2019	10/15/2019	10/15/2019		
Group	*Species Accumulation* *Scientific Name*	*Common Name*														
Anthozoa	*Bunodosoma capense*	Sea Anemone	0	0	0	1	0	0	0	0	0	0	0	0	1	8.33
Platyhelminthes	*Procerodes sp.*	Flatworm	0	0	0	0	0	0	1	1	0	0	1	1	4	33.33
Nemertea	*Cerebratulus fuscus*	Nemertean	0	0	0	0	0	0	1	0	0	0	0	0	1	8.33
	Nemertean unidentified	Nemertean	0	0	0	1	0	0	0	0	0	0	0	0	1	8.33
Polycheata	*Arabella iricolor*	Polychaete	0	0	0	0	0	0	0	0	0	1	0	0	1	8.33
	Perinereis namibia	Polychaete	0	0	0	1	0	0	0	0	0	1	1	1	4	33.33
	Pseudonereis variegata	Polychaete	0	0	0	0	0	0	0	0	0	1	0	0	1	8.33
	Family: Syllidae	Polychaete	0	0	0	0	0	0	0	0	0	1	0	0	1	8.33
Arachnida	*Amaurobioides africana*	Spider	1	1	0	0	0	0	1	0	1	0	0	0	4	33.33
Chilopoda	*Geophilomorpha sp.*	Centipede	1	0	0	0	0	0	0	0	1	0	0	0	2	16.67
Insecta	*Anurida maritima*	Collembolan	0	0	0	0	0	0	1	0	0	0	1	0	2	16.67
	Cercyon maritimus	Beetle	0	0	0	0	1	0	0	0	0	0	0	0	1	8.33
	Fucellia capensis	Fly	0	0	0	0	0	0	1	0	0	0	0	0	1	8.33

Class	Species	Common Name	1	2	3	4	5	6	7	8	9	10	11	12	#	%
	Stratiomyidae larvae	Fly Larvae	0	0	0	0	1	0	0	1	0	0	0	0	2	16.67
	Family: Staphylinidae	Beetle	1	0	0	0	0	0	0	0	0	0	0	0	1	8.33
	Family: Tabanidae	Fly Larvae	0	0	0	0	0	0	0	0	1	0	0	0	1	8.33
Isopoda	*Deto echinata*	Isopod	0	1	0	0	0	0	0	0	0	0	1	1	3	25.00
	Exosphaeroma truncatitelson	Isopod	0	0	0	0	0	0	0	0	0	1	0	0	1	8.33
	Ligia dilatata	Isopod	1	1	0	1	1	1	0	0	0	0	1	1	7	58.33
	Marioniscus spatulifrons	Isopod	0	1	0	0	0	0	0	0	0	0	0	0	1	8.33
	Philoscia sp.	Isopod	0	0	0	0	0	0	1	0	0	0	0	0	1	8.33
Amphipoda	*Eorchestia dassenensis*	Amphipod	0	0	0	1	1	0	0	0	0	0	0	1	3	25.00
	Paramoera bidentata	Amphipod	0	0	0	0	0	0	0	0	0	1	0	0	1	8.33
	Paramoera capensis	Amphipod	0	1	0	0	0	1	0	0	0	1	1	1	5	41.67
	Ptilohyale plumulosa	Amphipod	0	1	0	0	0	0	0	1	0	0	1	0	3	25.00
	Talorchestia capensis	Amphipod	0	1	0	0	1	1	0	1	1	0	0	0	4	41.67
Bivalvia	*Aulacomya atra*	Bivalve	0	0	0	0	0	0	0	0	0	1	0	0	1	8.33
Polyplacophora	*Acanthochitona garnoti*	Chiton	0	0	0	1	0	0	0	0	0	0	0	0	1	8.33
Gastropoda	*Burnupena cincta*	Welk	0	0	0	1	0	0	0	0	0	0	0	0	1	8.33
	Burnupena lagenaria	Welk	0	0	0	1	0	0	0	0	0	0	0	0	1	8.33
	Cymbula miniata	Limpet	0	1	0	0	0	0	0	0	0	0	0	0	1	8.33
	Cymbula oculus	Limpet	0	1	0	0	0	0	0	0	0	0	0	0	1	8.33
	Helcion pectunculus	Limpet	0	0	0	1	0	0	0	0	0	0	0	0	1	8.33
	Helcion pruinosus	Limpet	0	0	0	1	0	0	0	0	0	1	0	0	2	16.67
	Myosotella myosotis	Snail	0	0	0	1	0	0	0	1	0	0	1	0	3	25.00
	Oxystele antoni	Winkle	0	0	0	1	0	0	0	0	0	0	0	0	1	8.33
	Oxystele tigrina	Winkle	0	0	0	1	0	0	0	0	0	0	0	0	1	8.33
	Scutellastra longicosta	Limpet	0	0	0	1	0	0	0	0	0	0	0	0	1	8.33
Echinodermata	*Parvulastra exigua*	Sea Star	0	0	0	1	0	0	0	0	0	0	0	0	1	8.33
TOTAL Species Richness (#) per Site			4	9	0	15	5	3	6	5	4	9	8	6		

Appendix D Mean biomass (g/m²) per taxa across each of the 12 sites sampled for chapter 2 and totaled per site. Calculated as average biomass (g/m²) with standard deviation and average biomass percent (%) per taxa and totaled for the overall study.

Group	Scientific Name	Common Name	Blouberg 4/7/2019	Miller's Point 4/8/2019	V&A 5/17/2019	Kogelbaai 5/17/2019	Hermanus 6/17/2019	Grotto Bay South 08/13/2019	Blouberg South 09/13/2019	Grotto Bay North 09/16/2019	Grotto Bay 09/16/2019	Ganzekraal North 09/30/2019	Ganzekraal 10/15/2019	Ganzekraal South 10/15/2019	Average Biomass (g/m²)	Average Biomass %	Standard Deviation (+/- g/m²)
Species Accumulation																	
Anthozoa	*Bunodosoma capense*	Sea Anemone	0	0	0	29.75	0	0	0	0	0	0	0	0	2.48	0.76	214.70
Platyhelminthes	*Procerodes sp.*	Flatworm	0	0	0	0	0	0	0.5	1	0	0	0.01	0.02	0.13	0.04	7.74
Nemertea	*Cerebratulus fuscus*	Nemertean	0	0	0	0	0	0	1.5	0	0	0	0	0	0.13	0.04	10.83
	Nemertean unidentified	Nemertean	0	0	0	2.5	0	0	0	0	0	0	0	0	0.21	0.06	18.04
Polycheata	*Arabella iricolor*	Polychaete	0	0	0	0	0	0	0	0	0	40.5	0	0	3.38	1.04	292.28
	Perinereis namibia	Polychaete	0	0	0	0.25	0	0	0	0	0	20.75	2.91	4.31	2.35	0.72	149.11
	Pseudoereis variegata	Polychaete	0	0	0	0	0	0	0	0	0	11.5	0	0	0.96	0.30	82.99
	Family: Syllidae	Polychaete	0	0	0	0	0	0	0	0	0	0.25	0	0	0.02	0.01	1.80
Arachnida	*Amaurobioides africana*	Spider	0.5	0.5	0	0	0	0	0.25	0	0.5	0	0	0	0.15	0.04	5.63
Chilopoda	*Geophilomorpha sp*	Centipede	0.25	0	0	0	0	0	0	0	0.25	0	0	0	0.04	0.01	2.43
Insecta	*Anurida maritima*	Collembolan	0	0	0	0	0	0	0.75	0	0	0	0.03	0	0.07	0.02	5.40
	Cercyon maritimus	Beetle	0	0	0	0	0.25	0	0	0	0	0	0	0	0.02	0.01	1.80
	Fucellia capensis	Fly	0	0	0	0	0	0	0.75	0	0	0	0	0	0.06	0.02	5.41

Table heading: **Biomass (g/m²) per Site Total** — Sites — Sample date — Sites

Group	Species	Common name													Avg	%	Total
	Stratiomyidae larvae	Fly Larvae	0	0	0	0	1	0	0	0.25	0	0	0	0	0.10	0.03	7.28
	Family: Staphylimidae	Beetle	0.25	0	0	0	0	0	0	0	0	0	0	0	0.02	0.01	1.80
	Family: Tabanidae	Fly Larvae	0	0	0	0	0	0	0	0	8.75	0	0	0	0.73	0.22	63.15
Isopoda	*Deto echinata*	Isopod	0	188.75	0	0	0	0	0	0	0	0	1.47	0.01	15.85	4.88	1361.26
	Exosphaeroma truncatitelson	Isopod	0	0	0	0	0	0	0	0	0	0.5	0	0	0.04	0.01	3.61
	Ligia dilatata	Isopod	2	1083.25	0	7.25	484.25	9.25	0	0	0	0	0.01	0.26	132.19	40.72	8251.66
	Marioniscus spatulifrons	Isopod	0	323.25	0	0	0	0	0	0	0	0	0	0	26.94	8.30	2332.86
	Philoscia sp.	Isopod	0	0	0	0	0	0	0.25	0	0	0	0	0	0.02	0.01	1.80
Amphipoda	*Eorchestia dassenensis*	Amphipod	0	0	0	1	5.25	0	0	0	0	0	0	0.5	0.56	0.17	37.70
	Paramoera bidentata	Amphipod	0	0	0	0	0	0	0	0	0	1.5	0	0	0.13	0.04	10.83
	Paramoera capensis	Amphipod	0	0.5	0	0	0	0.25	0	0	0	0.5	0.05	0.48	0.15	0.05	5.49
	Ptilohyale plumulosa	Amphipod	0	0.75	0	0	0	0	0	0.5	0	0	0.12	0	0.11	0.04	6.17
	Talorchestia capensis	Amphipod	0	0.75	0	0	12.25	36.5	0	1.75	13.5	0	0	0	5.40	1.66	273.74
Bivalvia	*Aulacomya atra*	Bivlave	0	0	0	0	0	0	0	0	0	107.75	0	0	8.98	2.77	777.62
Polyplacophora	*Acanthochitona garnoti*	Chiton	0	0	0	34.75	0	0	0	0	0	0	0	0	2.90	0.89	250.79
Gastropoda	*Burnupena cincta*	Welk	0	0	0	152.25	0	0	0	0	0	0	0	0	12.69	3.91	1098.77
	Burnupena lagenaria	Welk	0	0	0	114.5	0	0	0	0	0	0	0	0	9.54	2.94	826.33
	Cymbula miniata	Limpet	0	199	0	0	0	0	0	0	0	0	0	0	16.58	5.11	1436.16
	Cymbula oculus	Limpet	0	14.5	0	0	0	0	0	0	0	0	0	0	1.21	0.37	104.64
	Helcion pectunculus	Limpet	0	0	0	2.75	0	0	0	0	0	0	0	0	0.23	0.07	19.85
	Helcion pruinosus	Limpet	0	0	0	206.5	0	0	0	0	0	11.5	0	0	18.17	5.60	1485.04
	Myosotella myosotis	Snail	0	0	0	306.5	0	0	0	24.75	0	0	0.28	0	27.63	8.51	2202.73
	Oxystele antoni	Winkle	0	0	0	213.75	0	0	0	0	0	0	0	0	17.81	5.49	1542.61
	Oxystele tigrina	Winkle	0	0	0	118.25	0	0	0	0	0	0	0	0	9.85	3.04	853.40
	Scutellastra longicosta	Limpet	0	0	0	1.25	0	0	0	0	0	0	0	0	0.10	0.03	9.02
Echinodermata	*Parvulastra exigua*	Sea Star	0	0	0	80.25	0	0	0	0	0	0	0	0	6.69	2.06	579.15
	TOTAL Biomass (g/m²) per Site		3.00	1811.25	0	1271.50	503.00	46.00	4.00	28.25	23.00	194.75	4.88	5.58	324.60	100.00	14938.15

Appendix E Mean density (#/m²) per taxa across each of the 12 sites sampled for chapter 2 and totaled per site. Calculated as average density (#/m²) with standard deviation and average biomass percent (%) per taxa and totaled for the overall study.

Group	Scientific Name	Common Name	Blouberg 4/7/2019	Miller's Point 4/8/2019	V&A 5/17/2019	Kogelbaai 5/17/2019	Hermanus 6/17/2019	Grotto Bay South 08/13/2019	Blouberg South 09/13/2019	Grotto Bay North 09/16/2019	Grotto Bay 09/16/2019	Ganzekraal North 09/30/2019	Ganzekraal 10/15/2019	Gan zekraal South 10/15/2019	Average Abundance (#/m²)	Average Abundance %	Standard Deviation (+/- #/m²)
Anthozoa	*Bunodosoma capense*	Sea Anemone	0	0	0	25	0	0	0	0	0	0	0	0	2.08	0.03	180.42
Platyhelminthes	*Procerodes sp.*	Flatworm	0	0	0	0	0	0	650	575	0	0	1	3	102.42	1.42	5969.93
Nemertea	*Cerebratulus fuscus*	Nemertean	0	0	0	0	0	0	25	0	0	0	0	0	2.08	0.03	180.42
	Nemertean unidentified	Nemertean	0	0	0	75	0	0	0	0	0	0	0	0	6.25	0.09	541.27
Polycheata	*Arabella iricolor*	Polychaete	0	0	0	0	0	0	0	0	0	50	0	0	4.17	0.06	360.84
	Perinereis namibia	Polychaete	0	0	0	50	0	0	0	0	0	750	467	690	163.08	2.26	7308.29
	Pseudonereis variegata	Polychaete	0	0	0	0	0	0	0	0	0	25	0	0	2.08	0.03	180.42
	Family: Syllidae	Polychaete	0	0	0	0	0	0	0	0	0	50	0	0	4.17	0.06	360.84
Arachnida	*Amaurobioides africana*	Spider	100	100	0	0	0	0	25	0	25	0	0	0	20.83	0.29	954.70
Chilopoda	*Geophilomorpha sp.*	Centipede	25	0	0	0	0	0	0	0	50	0	0	0	6.25	0.09	388.49
Insecta	*Amurida maritima*	Collembolan	0	0	0	0	0	0	775	0	0	0	6	0	65.08	0.90	5589.31
	Cercyon maritimus	Beetle	0	0	0	0	25	0	0	0	0	0	0	0	2.08	0.03	180.42
	Fucellia capensis	Fly	0	0	0	0	0	0	100	0	0	0	0	0	8.33	0.12	721.69
	Stratiomyidae larvae	Fly Larvae	0	0	0	0	150	0	0	25	0	0	0	0	14.58	0.20	1081.16

															Density (#/m²)	%	Total
	Family: Staphylinidae	Beetle	50	0	0	0	0	0	0	0	0	0	0	0	4.17	0.06	360.84
	Family: Tabanidae	Fly Larvae	0	0	0	0	0	0	0	0	25	0	0	0	2.08	0.03	180.42
Isopoda	*Deto echinata*	Isopod	0	1,400	0	0	0	0	0	0	0	0	21	11	119.33	1.66	10083.96
	Exosphaeroma truncatitelson	Isopod	0	0	0	0	0	0	0	0	0	125	0	0	10.42	0.14	902.11
	Ligia dilatata	Isopod	325	43,125	0	450	9,875	1,100	0	0	0	0	2	12	4574.08	63.47	311496.81
	Marioniscus spatulifrons	Isopod	0	3,600	0	0	0	0	0	0	0	0	0	0	300.00	4.16	25980.76
	Philoscia sp.	Isopod	0	0	0	0	0	0	150	0	0	0	0	0	12.50	0.17	1082.53
Amphipoda	*Eorchestia dassenensis*	Amphipod	0	0	0	50	625	0	0	0	0	0	0	89	63.67	0.88	4474.71
	Paramoera bidentata	Amphipod	0	0	0	0	0	0	0	0	0	275	0	0	22.92	0.32	1984.64
	Paramoera capensis	Amphipod	0	125	0	0	0	25	0	0	0	50	9	125	27.83	0.39	1195.63
	Ptilohyale plumulosa	Amphipod	0	75	0	0	0	0	0	75	0	0	26	0	14.67	0.20	728.66
	Talorchestia capensis	Amphipod	0	150	0	0	1,225	2,675	0	50	275	0	0	0	364.58	5.06	20172.30
Bivalvia	*Aulacomya atra*	Bivalve	0	0	0	0	0	0	0	0	0	25	0	0	2.08	0.03	180.42
Polyplacophora	*Acanthochitona garnoti*	Chiton	0	0	0	50	0	0	0	0	0	0	0	0	4.17	0.06	360.84
Gastropoda	*Burnupena cincta*	Welk	0	0	0	25	0	0	0	0	0	0	0	0	2.08	0.03	180.42
	Burnupena lagenaria	Welk	0	0	0	25	0	0	0	0	0	0	0	0	2.08	0.03	180.42
	Cymbula miniata	Limpet	0	25	0	0	0	0	0	0	0	0	0	0	2.08	0.03	180.42
	Cymbula oculus	Limpet	0	25	0	0	0	0	0	0	0	0	0	0	2.08	0.03	180.42
	Helcion pectunculus	Limpet	0	0	0	25	0	0	0	0	0	0	0	0	2.08	0.03	180.42
	Helcion pruinosus	Limpet	0	0	0	525	0	0	0	0	0	75	0	0	50.00	0.69	3778.30
	Myosotella myosotis	Snail	0	0	0	13,000	0	0	0	1,050	0	0	21	0	1172.58	16.27	93420.94
	Oxystele antoni	Winkle	0	0	0	325	0	0	0	0	0	0	0	0	27.08	0.38	2345.49
	Oxystele tigrina	Winkle	0	0	0	25	0	0	0	0	0	0	0	0	2.08	0.03	180.42
	Scutellastra longicosta	Limpet	0	0	0	25	0	0	0	0	0	0	0	0	2.08	0.03	180.42
Echinodermata	*Parvulastra exigua*	Sea Star	0	0	0	200	0	0	0	0	0	0	0	0	16.67	0.23	1443.38
	TOTAL Density (#/m²) per Site		500	48,625	0	14,875	11,900	3,800	1,725	1,775	375	1,425	553	930	7206.92	100.00	347731.99